Cognitive Radio Receiver Front-Ends

Analog Circuits and Signal Processing

Series Editors

Mohammed Ismail
Mohamad Sawan

For further volumes:
http://www.springer.com/series/7381

Bodhisatwa Sadhu • Ramesh Harjani

Cognitive Radio Receiver Front-Ends

RF/Analog Circuit Techniques

 Springer

Bodhisatwa Sadhu
IBM T. J. Watson Research Center
White Plains
New York
USA

Ramesh Harjani
Department of ECE
University of Minnesota
Minneapolis
Minnesota
USA

ISSN 1872-082X   ISSN 2197-1854 (electronic)
ISBN 978-1-4614-9295-5 ISBN 978-1-4614-9296-2 (eBook)
DOI 10.1007/978-1-4614-9296-2
Springer New York Heidelberg Dordrecht London

Library of Congress Control Number: 2013950224

Printed on acid-free paper

Springer is part of Springer Science+Business Media (www.springer.com)

Contents

Chapter 1
Introduction

1.1 Wireless Growth

Wireless technology has been evolving at a breakneck speed. The total number of cell-phones in use (as of 2011) was over 6 billion for a 7 billion world population [1] constituting 87 % of the world population. Additionally, with user convenience becoming paramount, more and more functions are being implemented wirelessly. For example, the U.S. army utilizes 40 different types of radios for its communications. Moreover, there is a considerable effort toward integrating all this wireless functionality in a single device. Smartphones today use as many as a dozen independent radios inside them.

1.2 Spectral Congestion

Unlike wired communications which use dedicated connections, wireless communications use particular frequencies of electromagnetic waves in space, thus sharing a common connection medium. Consequently, with the growth of wireless technology, spectral congestion has become a substantial concern, and threatens further growth of the technology [2].

However, this spectral congestion is a result of sub-optimal frequency utilization arising from a rigid spectrum licensing process. One example is spectrum licensing by the Federal Communications Commission, USA [3]. Currently, each frequency is strictly allocated for a particular application (for e.g. 850 MHz & 1.9 GHz for cell phones in the USA). When this application experiences excessive usage, the corresponding frequency becomes congested (causing dropped calls and busy lines for cell-phone users for instance). An example of frequency usage snapshot taken at Berkeley, California, USA [4] clearly shows heavy usage and congestion at a few frequencies (e.g. 0 to 1, 1.9, 2.4 GHz) and sparse usage at others (e.g. 0.5 % utilization in the 3–6 GHz band).

This inefficiency in frequency allocation can be resolved by allowing the unlicensed utilization of spectrum. However, this will necessitate new spectrum sharing

B. Sadhu, R. Harjani, *Cognitive Radio Receiver Front-Ends,*
Analog Circuits and Signal Processing 115, DOI 10.1007/978-1-4614-9296-2_1,
© Springer Science+Business Media New York 2014

Fig. 1.1 Concept of DSA

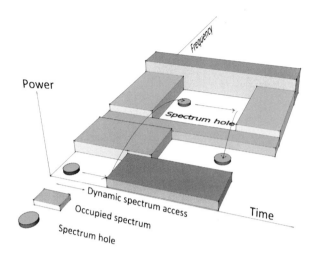

protocols to be developed in order to allow multiple users to utilize a single spectrum without causing harmful interference. A couple of approaches to spectrum sharing are being pursued: ultra-wideband (UWB), using a spectral underlay approach, and cognitive radio (CR), using a spectral overlay approach. In the ultra-wideband scenario (IEEE 802.15.3a, IEEE 802.15.4a), a secondary user is allowed to transmit in occupied bands, but with a power spectral density (PSD) so low that it is not deemed as harmful interference. The low PSD is compensated by the usage of a very wide bandwidth (> 500 MHz) to allow a significant transmit rate. In the cognitive radio scenario (IEEE 802.22), dynamic spectrum access (DSA) is utilized for spectrum sharing. This work focuses on the cognitive radio scenario for efficient spectrum utilization.

1.3 Dynamic Spectrum Access

In dynamic spectrum access, a secondary user is allowed to utilize other allocated but temporarily unused frequencies, to solve the problem of spectral congestion. Figure 1.1 shows a diagrammatic representation of the concept. An energy detector detects the power levels at different frequencies on a real-time basis (*spectrum sensing*) and determines spectral occupancy. The different colored cuboids in the figure represent spectral occupancy by different protocols, with different bandwidths, at different power levels, and for varying amounts of time. Frequencies not being used temporarily give rise to spectrum holes as shown using green discs. Dynamic spectrum access strives to seek out these spectral holes, and operate at these unused frequencies. This can greatly improve spectrum usage efficiency, and drastically reduce the problem of spectral congestion. For example, if the users of the 1.88–1.9 GHz frequency (cordless phones) are allowed to use the higher under-utilized frequencies, congestion in these devices could be avoided. In fact, the different spectrum governing bodies around the world are exploring new standards to allow this kind of communication (IEEE 802.22 in the USA).

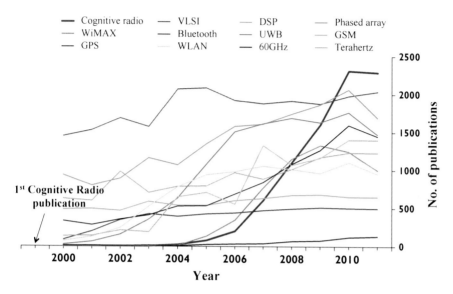

Fig. 1.2 Number of publications on different research areas in the last decade

1.4   Cognitive Radio

The use of dynamic spectrum access to intelligently improve communication efficiency has been the driving force behind the cognitive radio concept. Unfortunately, the definition and scope of what constitutes a cognitive radio has remained unclear to this day (an example definition by the FCC: [5]). In the narrow sense, a cognitive radio is an artificially intelligent device that can dynamically adapt and negotiate wireless frequencies and communication protocols for efficient communications[1]. For this, each participating device will need to be extremely flexible and capable of performing at different frequencies with their different regulations and requirements. Such a radio should have capabilities such as: determination of location, sensing the spectrum used by neighboring devices and analyzing the external spectral environment, to altering frequency and bandwidth, output power level adjustment, and even altering transmission parameters and protocols [6]. Interestingly though, if these devices are capable of such functionality, they would also be able to function as multiple wireless radios and therefore function as an integrated and complete wireless solution. For example, the same wireless radio could work as a cell phone, a GPS receiver, a garage door opener, and even as a remote control for other devices as shown in Fig. 1.3. Note that this is different from multiple different radios *assembled* in a smart-phone.

Figure 1.2 provides an indication of the growth of cognitive radio technology as a research area since its inception. The figure shows the number of publications

---

[1] It is envisioned, that in the future, the cognitive radio functionality will also include spacio-temporal intelligence, and significant learning capabilities heralding a new era in wireless communications.

Fig. 1.3 Functionality of an envisioned cognitive radio

with different keywords published per year in IEEE publications. Many keywords represent growing research areas in wireless communications, while other popular keywords such as 'VLSI' and 'DSP' have been included for comparison. The first cognitive radio paper was published in 1999; however, research in this area was relatively dormant till 2006. Since then, cognitive radios have seen a tremendous growth in research activity, and is now one of the most researched areas in wireless.

A cognitive radio can be separated into (1) a software defined radio (SDR) unit that forms the hardware of the cognitive radio, and an intelligence unit, that provides the required software based intelligence (cognition) to the radio. In this book, the SDR unit, and more specifically, the receiver front-end of the SDR, will be emphasized.

Like any communication device, the SDR needs a transmitter and a receiver. However, unlike a traditional wireless tranceiver, a cognitive radio not only needs to perform signal transmission and reception (*signaling*) but also needs to monitor the spectrum in real-time (*spectrum sensing*) in order to execute dynamic spectrum access. Spectrum sensing provides a wideband frequency snapshot helping the cognitive radio identify currently unused (potentially usable) frequencies. The cognitive radio then decides which unused frequency and protocol to use, and starts signaling at that frequency [7] (Fig. 1.3).

1.4.1 Spectrum Sensing

The dynamic spectrum access relies on dynamic spectrum monitoring using a spectrum sensing device. Among other features, this continuous wideband monitoring makes the cognitive radio hardware unique. From the hardware perspective, the spectrum sensor remains extremely challenging. Even for narrowband (small frequency range, < 100 MHz) spectrum sensing, limiting the power consumption is a challenge. On the contrary, the cognitive radio spectrum sensor needs to detect signals at all frequencies of interest instantaneously. Additionally, it needs to detect signals more unfailingly (about two orders of magnitude better sensitivity) than narrowband receivers to overcome the hidden-terminal problem (Fig. 1.4), shadowing, channel fading, multi-path, etc., lest it causes interference to other users due to incorrect sensing [4].

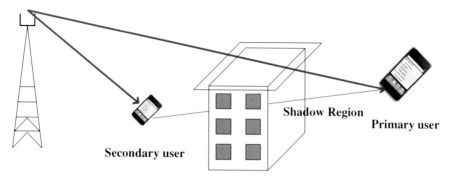

Fig. 1.4 The hidden terminal problem in cognitive radios

1.4.2 Signaling

The signaling in a cognitive radio is somewhat similar to that in traditional radios, in the sense of it being relatively narrowband. However, the cognitive radio signaling transceiver needs to be far more flexible. It should have the capability of operating over a very wide frequency range, changing its bandwidth of operation, altering the transmitted power, as well as using multiple standards for communications.

1.5 Organization

This book focuses on the architecture and circuit design for cognitive radio receiver front-ends.

Chapter 2 explores the different types of architectures for signaling and sensing in software defined radios for cognitive radio applications. A number of competing architectures are reviewed and discussed. Individual blocks and circuit requirements for these architectures are identified, and candidate implementations in literature are described. For spectrum sensing, the need for analog signal processing prior to digitization to lower the total power consumption, is emphasized. Two circuit blocks: the wide-tuning frequency synthesizer for signaling, and the analog signal processor for sensing, are identified as particularly challenging circuit blocks for SDR realization. These circuit blocks are discussed in later sections.

Chapter 3 discusses the design of wide tuning range, low phase noise VCOs for use in SDR signaling applications. The challenges to wide tuning range alongside low phase noise and superior power performance are discussed, followed by an identification of a viable solution using inductor switching. Two prototype designs are discussed that provide excellent power and phase noise performance over a very wide tuning range. Simulation and measurement results of tuning range, power dissipation, and phase noise across the tuning range are presented.

Chapter 4 explores the concept of RF sampling followed by discrete time signal processing prior to digitization for spectrum sensing applications. Specifically, frequency domain discrimination techniques are emphasized to reduce digitizing power. An example architecture is utilized for system level comparison based on analytical power estimates.

Chapter 5 discusses the design of a discrete Fourier transform based RF signal processor prior to digitization. The concept, implementation details, and handling of non-idealities in a Charge Reuse Analog Fourier Transform (CRAFT) engine prototype are described. Measurement results are presented.

Chapter 6 summarizes the book, and draws several conclusions regarding the design of SDR based cognitive radios.

Chapter 2
Cognitive Radio Architectures

2.1 Introduction

The system architecture for the SDR analog/RF is significantly different from that of traditional narrowband radio systems. In the original software radio proposal by Joseph Mitola in 1992 [8], he proposed an architecture where in the receiver,the RF bandwidth is digitized (no down-conversion), and signal analysis and demodulation is performed in the digital domain. Similarly, in the transmitter, the RF signal is synthesized in the digital domain, converted to analog and transmitted. The conceptual transceiver architecture is shown in Fig. 2.1. The Mitola architecture provides the maximum amount of flexibility through an increase in software capability. However, this architecture imposes impractical requirements on the analog-to-digital and digital-to-analog converters necessary for this architecture. For example, as discussed in [9], a 12 GHz, 12-bit ADC that might be used in a Mitola receiver would dissipate 500W of power! As a result, the ideal goal of communication at any desirable frequency, bandwidth, modulation and data rate by simply invoking the appropriate software remains far from realizable.

A number of SDR[1] architectures for cognitive radios currently being pursued provide only a subset of the functionality of the original software radio proposal. Specifically, in a practical interpretation, a large number of features of the waveform are defined in software, while the SDR hardware provides re-configurability to alter the waveform within certain bounds defined by the actual system. This re-configurability is commonly expected to encompass at least multiple radio access technologies (RAT) using a single transceiver IC over a wide frequency bandwidth [10]. However, in this book, we will focus on a much broader scope of cognitive radios that does not limit itself only to multi-standard, multi-band communications.

In order to co-exist with currently employed transceivers without causing them harmful interference, the SDR often needs to incorporate the most stringent

---

[1] Note that a software-defined radio (SDR) is not the same as a software radio as prosposed by J. Mitola which relies on RF digitization. In this book, we reserve the term SDR for all radio architectures that provide adequate flexibility for cognitive radio functionality.

B. Sadhu, R. Harjani, *Cognitive Radio Receiver Front-Ends,*
Analog Circuits and Signal Processing 115, DOI 10.1007/978-1-4614-9296-2_2,
© Springer Science+Business Media New York 2014

Fig. 2.1 Functionality of an
envisioned software radio by
J. Mitola [8]

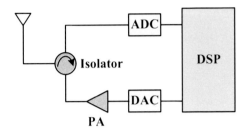

specifications among all the radio access technologies being employed in the
frequency range of interest.

For cognitive radio functionality, the SDR architecture can be considered a combination of a spectrum signaling transceiver system, and a sensing receiver system.
Architecture options for these two systems are discussed below.

2.2 Signaling

For purposes of signaling, the SDR is essentially narrowband. As observed in [9],
for most civilian applications of cognitive radios, only a particular narrow band of
RF frequencies are of interest at a particular time for transmission of the message
signal. However, this narrow frequency band may occur anywhere along the entire
bandwidth of operation of the SDR. In order to cover the entire range, multiple
receiver front-ends may be used, each dedicated to its narrow band of operation [11,
12]. However, it is easy to see that such an approach is power and area inefficient.
Instead, it is desirable to use a wide-tuning range frequency synthesizer in conjunction
with a wideband/wide-tuning[2] front-end for covering the frequency range as shown
for a receiver in Fig. 2.2. The number of such front-ends required will only be equal
to the maximum number of non-contiguous frequency channels to be simultaneously
used (typically assumed to be a single channel at present).

2.2.1 Receiver

An example wide-tuning receiver architecture is shown in Fig. 2.2. Each component
in this architecture is discussed below.

RF Bandpass Filters For the receiver architecture, co-existence with employed
technologies necessitates the use of RF bandpass filters. As shown in Fig. 2.2, these

---

[2] In this book we make the following distinction between wideband and wide-tuning circuits. We use
the term wideband for circuits that work over a large instantaneous bandwidth in RF. Wide-tuning
circuits, on the other hand, function over a narrow instantaneous bandwidth, but can be tuned over
a large frequency range.

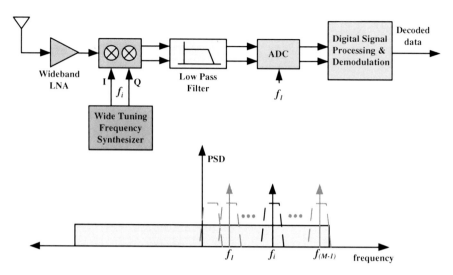

Fig. 2.2 A narrowband, wide-tuning signaling/scanning approach for signal reception and decoding

filters precede the LNA, and therefore their insertion loss directly compromises the receiver noise figure. Also, in order to remove large out-of-band blockers typical in the RF environment, high quality factors are mandated. Consequently, traditional RF architectures commonly use off-chip surface acoustic wave (SAW) filters to fulfill the front-end filtering requirements.

Unfortunately, SAW filters do not offer frequency tuning, and therefore, are not suitable for the SDR signaling architecture desired. As a result, the RF front-end filter is an important criteria for the realization of efficient SDRs. There has been significant research on tunable RF bandpass filters in recent years. A number of MEMS-based tunable filters, based on switched capacitors, have been developed [13–17]. A variety of varactors have also been utilized to realize frequency tuning [18–20].

More recently, driven by the vision of complete integration, a number of on-chip tuned filters are being developed. After initially working with the idea of on-chip switched capacitor LC filters [21], these filters have more recently been based on a revived idea originally published in 1960 [22] called the N-path filter. Recent work on this concept, using frequency-translation of complex impedances to a local oscillator frequency, is promising [23–27].

LNA Traditionally, LNAs have been designed as narrow-band solutions to attain a required performance over a small bandwidth. For the purpose of signaling in SDRs, these architectures need to be extended to one of three possibilities:

1. *Multiple narrowband LNAs in parallel:* As discussed already, this is a wasteful idea in terms of area; moreover, it requires the implementation of RF switches that are challenging to design, and increase parasitics in the RF path.
2. *A wideband LNA:* Wideband LNAs can be designed either by preceding a narrowband design with a wideband LC-ladder filter [28, 29], or using broadband

architectures [30]. The former uses multiple inductors using up valuable real-estate. The latter use MOS transistors and resistors (broadband elements) but typically suffer from a noise figure higher than 5 dB when power matched at the input. Such a high noise figure does not satisfy the stringent requirements of an SDR. Using global feedback improves the noise figure, but at the expense of stability [31]. Noise canceling schemes have been proposed to improve the noise figure below 2.5 dB for wideband architectures [9, 32], and are a promising candidate for SDRs.

However, wideband architectures provide little suppression for out-of-band blockers. Consequently, their broad input bandwidth increases the linearity requirements significantly.

3. *A wide-tuning narrowband LNA:* For signaling purposes, narrowband LNAs that are tunable are the preferred option. They not only provide suppression for out-of-band blockers but are expected to provide a lower noise figure (< 2 dB) compared to their broadband counterparts. However, tuned LNAs have so far either provided multi-band (discrete frequency switching) operation [33, 34] or low tuning range (e.g.: 23 % in [35]). For wider tuning range, switched inductors [36] may be utilized.

Frequency Synthesizer A wide-tuning range frequency synthesizer is a critical building block for the signaling receiver [37]. The following requirements can be identified for the synthesizer:

- Wide-tuning range, at least in excess of 67 % (f to $2 \times f$) such that lower frequencies can be obtained by integer division
- Low phase noise, such that the phase noise skirts do not cause down-conversion of spurious signals onto the desired signal (reciprocal mixing)
- Fast-settling behavior, to enable fast band-switching in the SDR
- Fine-frequency resolution determined by the minimum channel spacing desired

The first two requirements stem from the limitations in the voltage controlled oscillator (VCO). Unfortunately, these two performance features of VCOs: phase noise, and tuning range, strongly trade-off with one another. On-chip VCOs are based on two popular architectures: LC tank based, and ring based. Of these, LC tank VCOs are traditionally well suited for low phase noise applications, but suffer from a low tuning range. On the contrary, ring oscillators provide a wide tuning range, but suffer from poor phase noise. Consequently, there has been a significant effort toward realizing wide-tuning range LC VCOs for SDR applications.

Varactor based tuning typically provides 10–15 % tuning range [38, 39]. Use of switched capacitors in conjunction with varactors can provide a larger tuning range, but is still limited to about 50 % [40, 41]. Figure 2.3 shows the frequency tuning range and center frequency of published LC VCOs in the last decade. As seen, only one VCO (out of about 100) covers the tuning range required (66.67 %) to obtain all lower frequencies by division. Moreover, the tuning range requirements are made more challenging by PVT variation.

In Chap. 3, a new technique for obtaining a very wide-tuning range (up to 160 %) in LC VCOs while providing low phase noise throughout the tuning range is discussed.

Fig. 2.3 A survey of frequency tuning range of LC VCOs published in IEEE journals and conference proceedings appearing between 1992 and 2010

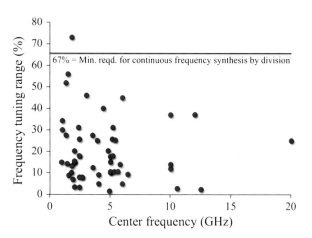

The oscillators described are based on switched inductors, and provide a viable option for the SDR signaling architecture.

Mixers For complete on-chip solutions, homodyne architectures have been the preferred choice. Homodyne architectures ease the requirements on the IF filter; however, they suffer from a number of issues including flicker noise, dc offsets, and second order non-linearity in circuits[3]. A high-pass filter may be used to filter out the low frequency components; however, this contributes to information loss and is not the preferred solution for very narrow-band architectures (e.g. GSM). Consequently, homodyne architectures have gained popularity for moderate to wide bandwidths, especially in SDR applications. Due to their lower flicker noise and excellent linearity, passive mixers have often been the preferred choice for SDR applications [9, 42].

Considering that a tuned RF filter does not provide as effective out-of-band rejection as fixed RF filters, and that SDR applications often require more stringent out-of-band rejection specifications [42], harmonic reject mixers become critical. To understand the problem of harmonics, consider the scenario shown in Fig. 2.4a. Despite the use of a variable bandwidth, band-select filter, large out-of-band blockers are not sufficiently suppressed. Subsequent mixing with a square wave (most mixers perform hard-switching) down-converts the desired signal, but also the interferers around the LO harmonics as shown in Fig. 2.4b. Consequently, the IF (zero-IF in this case) signal is significantly corrupted as shown in Fig. 2.4c.

To alleviate the problem of harmonic mixing, multi-phase harmonic rejection mixers can be employed [26]. Such mixers have been employed in a number of SDR architectures [9, 42], providing 30–40 dB harmonic rejection. Further harmonic rejection can be obtained using digital correction [42].

In addition, mixer first architectures [43], and a linear LNTA followed by a passive mixer [42] architectures have gained popularity due to their improved linearity and subsequent capacity to handle large out-of-band interference.

---

[3] In case a single mixer is used, it also suffers from LO pulling, LO re-radiation, reciprocal mixing, etc. However, using a multi-step down-conversion mitigates these effects.

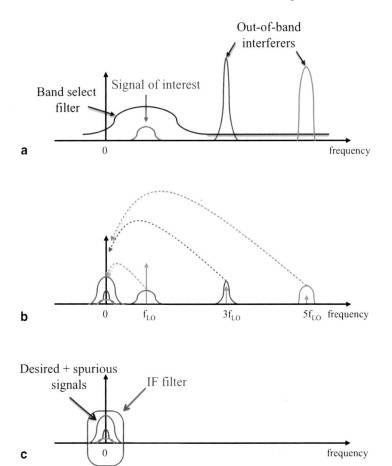

Fig. 2.4 The problem of harmonic down-conversion in a receiver

Baseband Filters Prior to digitization by an ADC, a variable bandwidth analog baseband filter is required to attenuate interferers to reduce the ADC dynamic range. Additionally, the baseband filter acts as a variable pole anti-aliasing filter (AAF) prior to sampling. Various techniques can be used to implement the baseband filter, broadly classified into continuous time and discrete time varieties. Continuous time filters have the advantage of not suffering from aliasing issues, but typically have poorer accuracy due to matching limitations. These filters can be realized using op-amps or using Gm-C filters [44–46]. Programmability can be achieved by switching resistors and capacitors, or by tuning these elements. However, the accuracy of these filters in sub-micron processes, and their lack of easy programmability have made way for active research in discrete time variations for SDR applications.

Discrete-time baseband filters are based on switched capacitors and can be implemented with or without active elements [9, 47–50]. The ease of programmability in these filters arises from the dependence of the filter notches on the clock frequency which can be tuned easily. Note that discrete-time filters require an anti-aliasing pre-filter that may be constructed using a combination of mixer output poles, current domain sampling, etc.

Analog to Digital Converter (ADC) ADCs for signaling applications typically require a high dynamic range to accommodate a variety of access technologies, and the remaining in-band blockers. Additionally, a relatively large (tens of MHz) bandwidth is required to accommodate wide-band access technologies.

For high dynamic range applications, $\Sigma - \Delta$ ADCs [51] have been the popular choice. $\Sigma - \Delta$ converters are based on $\Sigma - \Delta$ modulators that perform noise shaping on an oversampled signal using negative feedback techniques. The feedback loop suppresses the non-idealities in the quantizer, improving the effective resolution despite the use of a lower resolution quantizer. The shaped noise is filtered in the digital domain to obtain the ADC bits; as a result of the large amount of digital processing involved in the $\Sigma - \Delta$ ADC, these converters scale well with technology. Typically, $\Sigma - \Delta$ converters have been bandwidth limited due to quantizer latency, and parasitic poles, and sampling speeds have been limited to 1 GHz for 70 dB dynamic range. This results in bandwidths in the tens of MHz. However, in recent years, wideband $\Sigma - \Delta$ architectures that provide both the high dynamic range of $\Sigma - \Delta$ ADCs, as well as wideband digitization have been explored [52, 53]. These circumvent the speed limitations to provide signal bandwidths above 100 MHz alongside a high (> 70 dB) dynamic range.

For obtaining even larger bandwidths (usually at the expense of dynamic range), Nyquist ADCs are the more popular choice. Pipeline ADCs are the most popular choice for increasing the input bandwidth (> 200 MHz with 14–16 bit resolution) [54, 55]. For increasing the dynamic range, calibration techniques are popularly used [56]. However, the use of high-gain, wideband amplifiers and complex calibration techniques increase their power consumption, area, and design complexity.

Recently, successive approximation register (SAR) Nyquist ADCs have been used for moderate dynamic range (8–11 bit) at moderate speeds (< 100 MHz). SAR ADCs provide the advantage of very high power efficiency (low FOM) as they use just a single comparator for comparison [57–59]. However, they use N steps for providing an N-bit resolution, limiting their inherent speed. To overcome the speed limitation, time-interleaving is popularly used with the SAR architecture. Additionally, the SAR ADC resolution is limited by gain, offset, timing errors, and capacitor matching, and calibration techniques can be used for overcoming these [60].

A plot of the dynamic range and bandwidth of published ADCs [61] is shown in Fig. 2.5.

2.2.2 Transmitter

A generic transmitter architecture for SDR signaling applications is shown in Fig. 2.6. The critical blocks in this architecture comprise a wide-tuning frequency synthesizer, similar to one discussed in Sect. 2.2.1, a wideband mixer, similar to one discussed in

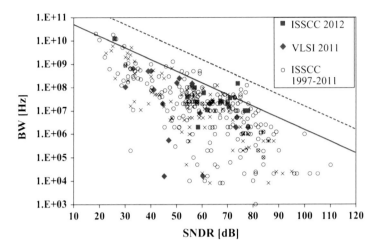

Fig. 2.5 A survey of ADCs from 1997–2012 [61]

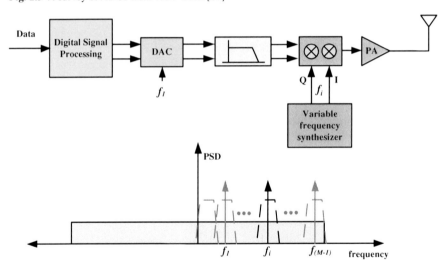

Fig. 2.6 A narrowband, wide-tuning approach for signal transmission

Sect. 2.2.1, and a wideband, linear power amplifier (PA). Linearization techniques for wideband SDR architectures are discussed in [62]. Harmonic mixing [26] may be used; moreover, the mixer and power amplifier may be combined to obtain a linear power mixer [63].

2.3 Spectrum Sensing

Spectrum intelligence or knowledge about the spectral and spatial electromagnetic spectrum around us is of critical importance for dynamic spectrum access in cognitive radios. One of the primary bottlenecks in implementing dynamic spectrum access is reliably detecting primary users before deciding on the frequency, channel, access and modulation formats of transmission for the secondary user. The problem of primary user detection is further complicated by multi-path fading and hidden node/shadowing effects [4]. Some of these problems can be mitigated somewhat by using cooperative techniques across multiple sensors [64]. However, the issue of identifying primary user activity with varying modulation formats and signal power levels, over a wide swathe of spectrum, using limited power continues to confound the industry.

From an SNR perspective, the optimal method for signal detection is a matched filter. However, such a coherent detection technique requires *a priori* knowledge of the primary user signals and modulation formats. Worse still, a cognitive radio based on coherent detection with matched filters would require a dedicated radio for each primary user class. One method to address this problem is to employ a blind, non-coherent detection scheme using energy detection. Despite the implementation simplicity of this approach, non-coherent energy detection schemes require a large number of samples $[O(1/SNR^2)]$ and therefore, a longer sensing time. Moreover, such a scheme would be unable to distinguish noise and in-band interference from primary user signals, or work for spread spectrum signals (direct sequence and frequency hopped). An intermediate approach would be to use feature detectors which rely on some knowledge (or feature) of the primary user. Cyclostationary feature detectors rely on detecting the built-in periodicity of modulated signals such as their carrier frequency, bit rate, repeated spreading, pulse strains, hopping sequences, and pilots. Moreover, even in the absence of clock timing or phase knowledge, cyclostationary feature detectors obtain considerable improvement in detection performance [65]. A brief description of the spectral correlation function on which cyclostationary feature detectors are based is provided in Appendix A.

Spectrum Sensing Architectures   There are two fundamentally distinct options for realizing a spectrum sensing receiver for an SDR front-end: a scanner type, and a wide instantaneous bandwidth digitizer type.

1. Scanner: In this scheme, a narrowband, wide-tuning receiver scans and digitizes the entire bandwidth (similar to a spectrum analyzer) progressively for digital analysis. The digital signal processing backend processes each digitized band sequentially and stitches the frequency domain outputs to obtain a spectral map of the environment. The architecture used for this scheme is very similar to that used in the signaling receiver discussed above (Fig. 2.2). However, in order to overcome practical wireless communications issues such as multi-path, fading, hidden nodes, interference problems in the context of an unknown signal, etc [4, 66], the sensitivity and dynamic range requirements of the architecture are more challenging than the signaling scheme. Moreover, note that sensing may be a blind detection problem, as opposed to signaling where *a priori* knowledge of the transmitted signal is available.

   Although the scanning architecture is able to re-use much of the signaling architecture (or vice versa), this detection technique suffers from multiple short-comings. These systems lack the agility required to be able to detect any fast-hopping

Fig. 2.7 A wideband RF to
digital conversion architecture
for spectrum sensing

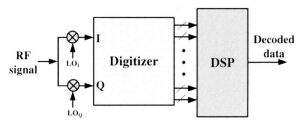

signals. Frequency domain stitching in the digital domain is power hungry due to the phase distortion of the analog filters that select each digitized sub-band. Moreover, stitching the frequency domain information from multiple scans is imperfect in the face of multi-path; consequently, signals spanning multiple scan bandwidths are imperfectly reconstructed. Due to these and other reasons, it is desirable to construct a real-time instantaneous bandwidth digitizer (similar to J. Mitola's original software radio idea) as part of the spectrum sensor.

2. Wide instantaneous band digitizer: Unlike the scanning type architecture, a wide instantaneous band digitizer digitizes the entire wide RF bandwidth instantly, and performs the digitization in a real-time. Understandably, the wideband digitizer has widely been considered as one of the bottlenecks to the realization of an SDR based cognitive radio. A significant number of efforts in recent years have focused on parts of this larger issue by addressing wider bandwidths, broadband matching, higher front-end linearity, and most importantly, wideband analog to digital converters.

Several architectures have been proposed for the RF front-end. Of these, the most popular is the generalization of the Mitola receiver architecture as shown in Fig. 2.7 effectively performing an RF to digital conversion (R-to-D). Typically, the front-end requires a wideband low noise amplifier prior to the RF digitizer. Moreover this front-end needs to handle a large dynamic range due to the large peak to average power ratio (PAPR) of broadband signals. The increase in the PAPR for larger bandwidths is depicted in Fig. 2.8. As shown in the specific example, the PAPR for the narrowband signals is only 2, while that for a wideband signal (5 times the bandwidth) with multiple signals, all having similar powers, is 10. In this case, the PAPR grows linearly with an increase in bandwidth. As a result of this increased PAPR of the wideband inputs, a very linear RF front-end is necessitated. The linearity requirements of the LNA have been addressed in [67–69]. Another approach comprising a low noise transconductance amplifier (LNTA) followed by mixers is discussed in [42]. Moreover, passive mixer first topologies have also been proposed for high IIP_3 performance [43].

The digitizer block shown in the figure is essentially an equivalent ADC with performance specifications beyond that achievable using state-of-the-art converters discussed in Sect. 2.2.1. This wideband digitizer can be implemented using multiple techniques, all based on some form of multiplexing in order to ease the requirements on the ADCs. A time-multiplexed wideband approach using time-interleaving as shown in Fig. 2.9 was proposed in [70]. This scheme reduces the

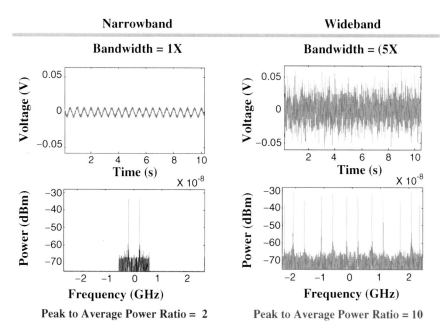

Fig. 2.8 PAPR increase in wideband signals compared to narrowband signals

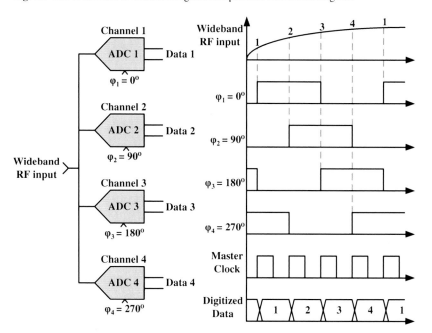

Fig. 2.9 Time interleaved ADCs for broadband channelization

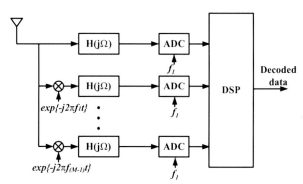

Fig. 2.10 Low pass filterbank approach for channelization

sampling rate of ADCs. However, all the ADCs continue to see the full input bandwidth, and therefore, still require high dynamic range capability. Also, the sample and hold circuitry remains difficult to design.

In order to reduce the dynamic range requirements on the ADCs, it is possible to transform the input signal to a different domain prior to digitization [71]. For example, a frequency domain transform is particularly attractive [72]. A frequency domain transform can be approximated by using band-pass filters for channelization, as proposed in [73]. This reduces the input dynamic range requirements of the ADCs that follow each individual bandpass output, but introduces the problem of designing impractically sharp band-pass filters. In [74], replacing sharp band-pass filters with frequency down-mixers followed by sharp low-pass filters eliminates this problem as shown in Fig. 2.10. However, this architecture is based on phase locked loops (PLL), mixers and low-pass filters [75], or on injection locked oscillators [76][4], and can be power hungry. Moreover, harmonic mixing of the input signal frequencies within the SDR input bandwidth severely corrupts the channelized baseband signals. Additionally, signal reconstruction from the digitized filter-bank outputs is challenging (Fig. 2.11).

In this book, we propose a digitizer approach based on analog signal processing using passive switched capacitors to condition the signal prior to digitization by ADCs (Fig. 4.1). The RF discrete time (DT) signal processing, as shown in the second block in Fig. 2.12, eases the dynamic range requirements on the ADCs by pre-filtering the signal.

As a specific example, we discuss in detail a digitizer approach based on frequency discrimination using an analog domain DFT followed by ADCs (Fig. 2.12). This technique enjoys the advantages of analog domain frequency discrimination: channelization and consequent ADC dynamic range reduction. In addition, it also enjoys an ultra-low power implementation, signal spreading benefits, no harmonic mixing problem, and simple reconstruction in the digital domain. Details on the architecture and implementation are included in Chap. 5.

---

[4] Note that injection locked oscillators have the advantage of a larger noise suppression bandwidth for the VCO ($\approx$ lock range) [77] and provide better reciprocal mixing robustness compared to PLLs (assuming the reference phase noise is better than the VCO phase noise).

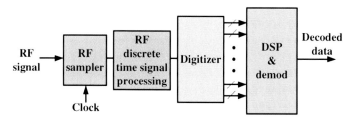

Fig. 2.11 An envisioned SDR architecture enabled by passive analog signal processing

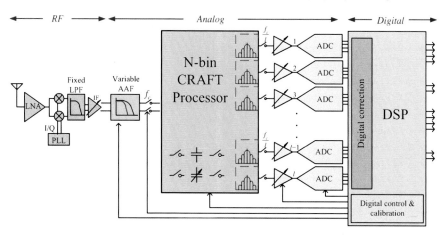

Fig. 2.12 An envisioned SDR architecture enabled by CRAFT

2.4 Conclusions

In this chapter, several software defined radio architectures for cognitive radio applications were discussed. The SDR system was divided into two functional blocks: signaling and sensing. Novel architectures for each of the blocks were reviewed. Additionally, new circuit topologies that are potential candidates for satisfying the required specifications were reviewed. Specifically, wideband, and wide-tuning range, and programmable circuits for RF and baseband filters, low noise amplifiers, mixers, frequency synthesizers, power amplifiers, and analog to digital and digital to analog circuits were discussed. Existing architectures for tackling the extremely challenging problem of wideband spectrum sensing were explored, and a new architecture for the same was proposed.

Chapter 3
Wideband Voltage Controlled Oscillator

3.1 Introduction

In this chapter, we explore techniques to increase the tuning-range of VCOs while maintaining adequate phase noise and power dissipation performance over the tuning range. Two popular oscillator architectures: LC tank based, and ring based, are considered for SDR signaling and spectrum scanning applications. Of these, LC tank VCOs are traditionally well suited for their superior phase noise and low power consumption at radio frequencies, and are therefore the preferred choice. However, LC tank VCOs are notorious for their lower tuning range as compared to ring oscillators. In this work, we select the LC tank oscillator and devise a scheme based on switched inductors, and capacitor array optimization, to extend the desirable power and phase noise properties of LC VCOs over a wide tuning range for use in SDR signaling (and sensing) applications.

3.2 Advantages of Switched-Inductor Tuning

The frequency of oscillation of an LC tank is given by:

$$f = \frac{1}{2\pi\sqrt{LC}} \tag{3.1}$$

Frequency variation can be realized by changing the capacitance and/or the inductance. For wide tuning ranges, traditional tuning by varying the capacitance with a fixed inductance is disadvantageous. For such requirements, switched-inductor resonators offer distinct advantages as described below.

Phase Noise The approximate phase noise of a resonator dominated by the inductor Q is derived from Leeson's model in (3.2), which for $\omega_0 \gg Q\Delta\omega$, reduces to (3.3).

$$L(\Delta\omega) = 10\log\left\{\frac{2\,FkT}{P_{sig}}\left(1+\left(\frac{\omega_0}{2Q\Delta\omega}\right)^2\right)\right\} \tag{3.2}$$

B. Sadhu, R. Harjani, *Cognitive Radio Receiver Front-Ends,*
Analog Circuits and Signal Processing 115, DOI 10.1007/978-1-4614-9296-2_3,
© Springer Science+Business Media New York 2014

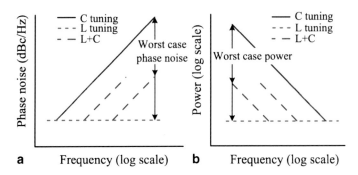

Fig. 3.1 VCO power & phase noise for alternate frequency tuning options

$$\approx \quad 10\log\left(\frac{kTFR_s}{V_{rms}^2}\left(\frac{\omega_0}{\Delta\omega}\right)^2\right) \tag{3.3}$$

Here $\Delta\omega$ is the frequency offset, F is the noise factor, R_s is the inductor's parasitic series resistance, P_{sig} is the signal power, ω_0 is the oscillation frequency, and Q is the inductor's quality factor. To the first order, R_s is proportional to L, and the expression further reduces to (3.4)

$$10\ \log\left(\frac{kTFAL}{V_{rms}^2}\left(\frac{\omega_0}{\Delta\omega}\right)^2\right) \tag{3.4}$$

where A is the constant of proportionality. Therefore, the phase noise becomes *directly proportional* to f^2L. For traditional LC VCOs with constant inductance and capacitive tuning, the phase noise is proportional to f^2. Consequently, even if the phase noise of the oscillator is carefully optimized for the lowest frequency, it is bound to degrade at the higher frequencies as shown in smooth blue in Fig. 3.1a.

However, from (3.1), for a constant capacitance, f^2L is a constant. Therefore, if the resonator can be tuned using inductance tuning with a constant capacitance, the phase noise would remain constant with frequency (to the first order) as shown in dotted green in Fig. 3.1a. Since inductance tuning is not realizable in bulk-CMOS processes, we use a combination of inductance switching and capacitance tuning to obtain a suitable compromise as shown in dashed red in Fig. 3.1a. Consequently, the worst case phase noise shows considerable improvement for switched-inductor resonators.

The improvement in phase noise by the introduction of a switch might seem counter-intuitive. Introducing a switch degrades the quality factor of an LC tank. However, as expressed in equation 3.4, the phase noise expression depends on the parasitic series resistance of the tank, rather than its quality factor. Therefore, as long as the parasitic resistance introduced by the switch is lower than the parasitic resistance of the inductor switched out, there is a net phase noise improvement due to the switching.

Power   Likewise, the power dissipation, for a constant voltage swing (V_{sw}), can be derived as,

$$V_{sw} = \frac{(\omega L)^2}{R_s} I_{bias} = \frac{(\omega L)^2}{R_s} \frac{Power}{V_{DD}}$$

$$\Rightarrow Power = \frac{R_s V_{sw} V_{DD}}{(\omega L)^2} \tag{3.5}$$

Again, to the first order, R_s is proportional to L, and the power dissipation becomes *inversely proportional* to $f^2 L$:

$$Power = \frac{A V_{sw} V_{DD}}{\omega^2 L} \tag{3.6}$$

where A is the constant of proportionality. For the traditional capacitive tuning scheme using a constant inductance, the power is inversely proportional to f^2. Consequently, even if the power dissipation is optimized at the highest frequency, it is bound to degrade at the lower frequencies as shown in smooth blue in Fig. 3.1b. Again, in the case of pure inductance tuning the power dissipation remains constant (to the first order) versus frequency as shown in dotted green in Fig. 3.1b. The tuning technique using switched inductance and variable capacitance provides a suitable compromise between these two methods with improved power dissipation, in comparison to the traditionally used fixed inductor, variable capacitance scheme, as shown in dashed red in Fig. 3.1b.

Tuning Range   The tuning range of capacitively tuned oscillators is limited by the parasitic capacitance at the highest frequency, and the startup criterion (3.7) at the lowest frequency.

$$g_m R_p \geq 1 \text{ from the startup condition}$$

$$\Rightarrow \quad gm \frac{\omega^2 L^2 R_s}{R_s^2} = gm \frac{L}{R_s C} \geq 1 \tag{3.7}$$

Switched-inductor resonators, on the other hand, use a larger inductance at lower frequencies enabling the startup criterion to be fulfilled over different frequency ranges for different inductance values. Effectively the total tuning range becomes the sum of the tuning ranges of each frequency bank obtained from switching the inductance values.

3.3   Prototype Design I

Based on the inductor switching idea described above, we now present a complete prototype design as a proof of concept. For this design, we used a frequency floorplan spanning 3.3–8.3 GHz and targeted a CMOS only implementation in the IBM 130 nm

Fig. 3.2 Cross-coupled
nMOS pMOS topology used
for the prototype design

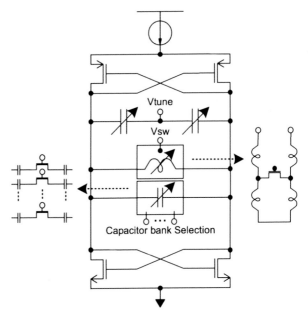

SiGe BiCMOS technology. For our particular application, since phase noise was an important criterion being considered, the cross-coupled nMOS pMOS topology in Fig. 3.2 was the appropriate choice.

For the design, an inductor switching scheme as proposed earlier was used. An octagonal inductor with the switch positioned within and underneath it was designed as shown in Fig. 3.3. Based on the design methodology provided in [78], for a 1.2 nH inductor, the switching ratio($k = L_{off}/L_{tot}$) and switch width were determined to be 0.55 and 500 μm respectively.

Within each frequency bank, the design was optimized based on the framework described in [79]. Special attention was paid to the layout of the VCO to ensure good tuning and phase noise performance. The effect of this is evident in the measurement results discussed later.

G_m **Cells** The W/L ratios of the g_m transistors were designed to allow a maximum current of 10 mA. This allows start-up at the lowest frequencies of interest in the two bands, but causes the oscillator to function in the current limited regime (non-optimal phase noise as discussed in [79]). Such a design technique offers two advantages: it reduces the parasitic capacitance offered by the g_m transistors, and it limits the worst-case power dissipation of the VCO. Although this causes a degraded phase noise performance at the lowest frequencies, the lower frequencies do not determine the overall worst case phase noise over the entire tuning range (Fig. 3.15). Therefore, this degradation does not affect the overall phase noise performance of the VCO. In effect, the improved phase noise at the lowest frequencies is traded off for obtaining an improved power performance and tuning range.

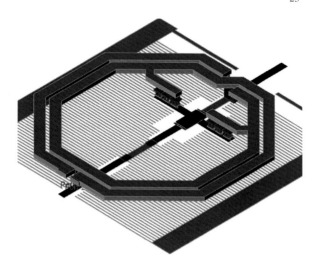

Fig. 3.3 3-D view of the switched inductor with the switch shown at a lower tier in the center of the inductor

Inductor The inductor was designed as shown in Fig. 3.3. For the switching ratio obtained, the inductor switch had to be physically connected between the arms of the inductor's inner coil. A very wide nMOS transistor (low resistance) was used as the switch, and was positioned directly beneath and within the inductor coil. As compared to placing the switch outside the spiral, this reduces the effects of interconnect parasitics significantly. Higher order electromagnetic(EM) effects of placing the transistor beneath the inductor were considered [80], and the switch was placed away from the traces in order to reduce unpredictable capacitive coupling effects. Also, in this switching scheme, the relatively lower quality inner part [81] of the inductor is switched out, as compared to previous designs [82], thereby promising a better phase noise performance. The large switch size adds parasitic capacitance to the LC tank. However, by placing the switch close to the midpoint (small signal ground) of the differential inductor, the effect of these parasitics is considerably reduced. A high resistance substrate beneath the inductor in conjunction with a patterned ground shield are used to improve the quality factor of the inductor.

For the purpose of circuit simulations, an S-parameter 4-port network was used to mimic the switched inductor. For this, EM simulations were performed using ADS Momentum®, and the resultant S-parameter data was exported to Cadence Virtuoso® design environment for circuit simulations. In Cadence Virtuoso®, a rational interpolation scheme was used for periodic steady state analysis. This provided a simple and efficient way to include EM effects for the custom-made inductor in circuit simulations.

Capacitor Bank For the capacitor bank, a 5-bit, binary-weighted, differential, switched, MIM capacitor array was constructed for coarse frequency tuning. For fine frequency tuning, two differentially connected MOS varactors were utilized. The combination of discrete and continuous tuning ensures a desirable, lower value

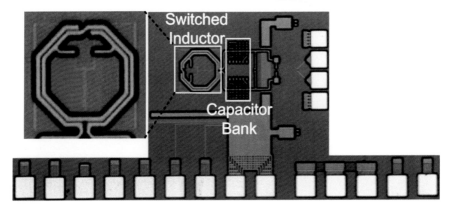

Fig. 3.4 Die photograph of wide tuning range VCO prototype

of K_{VCO} for the oscillator. The switched capacitor array was constructed using multiple unit cells to provide good matching and ensure a monotonic frequency tuning characteristic.

3.4 Measurement Results

The design was implemented in CMOS in the IBM 130 nm SiGe BiCMOS process. Figure 3.4 shows a die photograph of the VCO. The total area including the bondpads is 0.87 mm^2, and the area of the VCO core is 0.1 mm^2.

Measurement results for the prototype design are discussed below. A table comparing the performance of this VCO with other recent publications is shown in Table 3.2.

Tuning For measuring the tuning range, the leakage output was coupled out by an off chip antenna to avoid the effect of loading from the 50 Ω buffer and the probe pads. Using the combination of discrete and continuous tuning described above, the VCO achieves a frequency tuning range (FTR) of 87.2 %, from 3.28 to 8.35 GHz at room temperature. The FTR increases with a decrease in temperature to 88.6 % (3.28–8.52 GHz) at 0 °C. As seen in Table 3.2, the FTR obtained betters other previously reported wideband VCO solutions. When probed, the frequency tuning range reduces to 83.9 %, from 3.2 to 7.82 GHz, due to the loading by a sub-optimally designed probe buffer, but is still better than prior designs.

Figure 3.5 plots the frequency tuning of the VCO versus capacitor bank control voltages. Tuning in the low frequency band is shown using a + marked green line, and in the high frequency band using a * marked blue line. Note that all the binary weighted capacitors are not used for switching the frequency in the high frequency band. This is because, in the high band, the VCO fails the start-up condition for a lower capacitance as compared to that in the low frequency band.

Fig. 3.5 Measured tuning range versus capacitor bank control

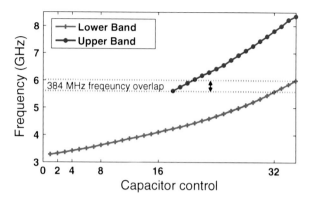

Fig. 3.6 Measured power dissipation versus frequency (log log plot)

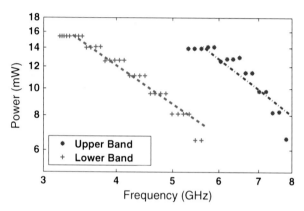

The two frequency bands obtained by inductor switching achieve a 384 MHz overlap as seen in Fig. 3.5. A considerable frequency overlap when switching capacitors is also ensured through the use of suitable varactors. This overdesign ensures a continuous frequency tuning in the face of process variations.

Power Dissipation The power requirement of the VCO, as predicted for a constant voltage swing, reduces as frequency increases in each individual band (Fig. 3.6). The measured power dissipation for the core VCO varies from 6.6 to 14.1 mW in the upper band, and 6.5 to 15.4 mW in the lower band from a 1.6 V supply. The trends expected in the variation of power dissipation with frequency are evident (compare with Fig. 3.16). The improvement in worst case power dissipation through inductor switching, as was expected from Sect. 3.2, is obtained.

Phase Noise For phase noise measurement, a buffer was included on-chip in order to drive a 50 Ω environment. The probed differential outputs were converted to a single-ended signal compatible with the measurement apparatus using an off-chip balun. Phase noise measurements were performed using an HP E4407B spectrum analyzer. A screenshot of the phase noise measurements at the lowest frequency point is shown in Fig. 3.7. The variation of the phase noise with the frequency (log scale)

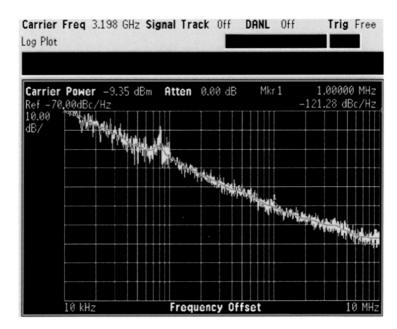

Fig. 3.7 Measured phase noise at 3.28 GHz

is shown in Fig. 3.8. The phase noise varies between -122 and -117.5 dBc/Hz at 1 MHz offset in the lower frequency band, and -119.6 and -117.2 dBc/Hz at 1 MHz offset in the higher frequency band. The trends expected in phase noise variation with frequency is evident (compare with Fig. 3.15). Again, an improvement in the worst case phase noise performance through inductor switching, as was expected from Sect. 3.2, is obtained.

In order to evaluate and compare VCO performances, the Power-Frequency-Tuning Normalized figure of merit (PFTN) as described in [79] was used. A temperature of 300 K was used for the PFTN calculations. Figure 3.9 shows the VCO performance over the entire frequency range. The PFTN varies between 6.6 and 10.2 dB in the low frequency band, and 6.7 and 9.5 dB in the high frequency band. As shown in Table 3.2, this design provides one of the best PFTN performances reported to date. Also notable is the lack of significant variation in the PFTN values over such a wide tuning bandwidth. This was ensured through a carefully designed trade-off between phase noise and power across the entire tuning range.

3.5   Design Extension

This section describes an extension of the design discussed in prototype I using the switched-inductor described in Sect. 3.3.

Fig. 3.8 Measured phase
noise versus frequency (log
scale)

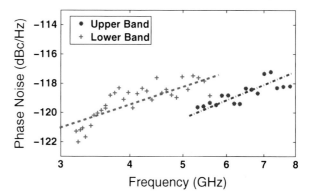

Fig. 3.9 PFTN versus
frequency (log scale)

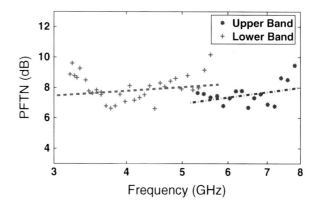

Topology In order to improve the phase noise performance, an nMOS-pMOS cross
coupled topology is used for the VCO design [79]. The VCO design for this topology
has been explored in [79, 83], and we base our design on the same framework. For
increasing the tuning range, we use a switched-inductor, and optimize the capacitor
bank as described in the sections below.

Switched-Inductor A switched-inductor was constructed and optimized similar
to the one described earlier. As shown in Fig. 3.10, two metal layers are used to
construct the inductor. For clarity purposes, a dual bank with a single nMOS transistor
switch is described here, although all the design concepts mentioned here can easily
be extrapolated for multiple banks with multiple switches. This switched-inductor
provides two frequency banks depending on the state of the switch. The low frequency
bank is obtained with the switch turned **off**; the high frequency bank is obtained with
the switch turned **on** shorting out a portion of the inductor and thereby lowering the
effective inductance.

Switched Capacitor Bank A switched capacitor bank is constructed using an 8-bit
binary weighted switched MIM capacitor array for discrete tuning, and a varactor for
continuous tuning. In the capacitor array, 5 bits are shared between the two frequency

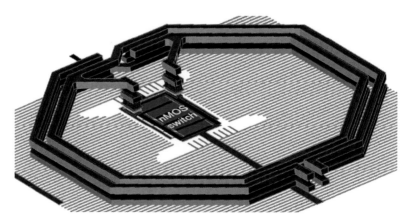

Fig. 3.10 A 3 dimensional view of the switched-inductor design

banks resulting from switching the inductor. The higher 3-bits (larger capacitance) do not allow the VCO to start up in the higher frequency bank (lower inductance) as expected from (3.7). These bits are used in the lower frequency bank (higher inductance) case only. The resultant tuning range of this VCO is therefore the sum of the tuning ranges of two individual resonators that use only capacitive tuning. The detailed design (with an emphasis on parasitics) of the capacitor array is critical to extend the tuning range of the VCO and is elaborated in the next section.

3.6   Detailed Capacitor Bank Design

Switch Sizing The MIM capacitors in the array are connected between the two differential VCO outputs 'Out+' and 'Out−' by nMOS transistor switches as shown in Fig. 3.11. The switches contribute parasitic capacitance when **off** reducing the tuning range of the oscillator. Also, they contribute parasitic resistance when **on** degrading the phase noise of the oscillator. A naive way of sizing these nMOS switches would be as follows: the switch for the smallest capacitor bit is sized so as to minimally degrade the phase noise at the highest frequency of operation (corresponding to the worst phase noise as shown in Fig. 3.1). The other switches are sized in proportion to the increasing capacitance to ensure a constant RC product (constant Q at a particular frequency) and a smooth tuning curve. However, for large capacitor banks, these proportionally sized switches contribute considerable parasitic capacitance. Since the Q of a capacitor is given by $Q = 1/\omega RC$, the capacitor Q improves at low frequencies. Therefore, the size of capacitor switches can be reduced for the larger capacitors which come into effect only at these lower frequencies. Such an implementation is shown in the layout of Fig. 3.12.

In effect, the Q of the capacitor bank is traded off with the parasitic capacitance of the switches, thereby improving the tuning range of the oscillator. Note also, that

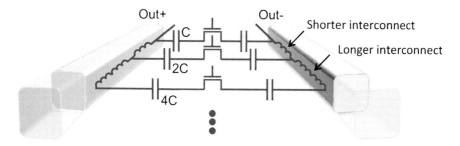

Fig. 3.11 Diagram showing the use of interconnect inductance to boost capacitances in an optimal ratio

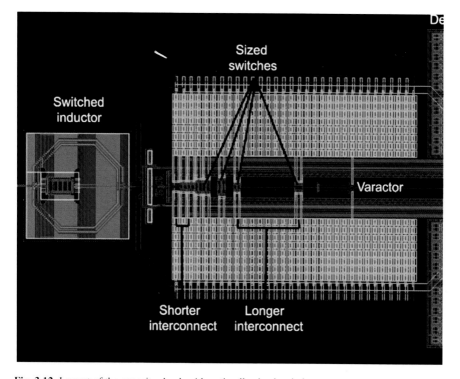

Fig. 3.12 Layout of the capacitor bank with optimally sized switches

since the quality factor of the inductor $Q = \omega L / R_s$ degrades at lower frequencies, the Q of the capacitor bank can actually be allowed to drop at the lower frequencies without affecting the overall resonator quality factor as given by $1/Q = 1/Q_{ind} + 1/Q_{cap}$. Again, in this particular design, since the three largest capacitor bits are turned on only in the low frequency bank, these switch sizes can be designed to be particularly small. For this design, the phase noise was kept approximately constant for every frequency at which a new capacitor was switched in. The switch sizes used

Table 3.1 Capacitor array switch sizing (lower frequency bank)

Cap. units	Max. freq.	Ind. Q	Switch size in normalized units		
			Constant RC	Constant cap. Q	Constant $L(\Delta\omega)$
1	3.8	15.5	1	1	1
2	3.7	15.4	2	2	2
4	3.6	15.3	4	4	4
8	3.3	14.8	8	7	7
16	3	14.2	16	15	11
32	2.5	13	32	27	16
64	2	11.3	64	51	11
128	1.4	8.6	128	88	15
$\Sigma = 255$			$\Sigma = 255$	$\Sigma = 195$	$\Sigma = 67$

were in the proportion 1:2:4:7:11:16:11:15 for binary weighted capacitors (sized in the proportion 1:2:4:8:16:32:64:128). This sizing is obtained through simulation such that, when a new capacitor is switched in, the phase noise is approximately equalized to the worst case phase noise. Consequently, the array switch contributed parasitic capacitance is reduced by 73.8 %, i.e., by a (255–67)/255 ratio. The technique is summarized for the lower frequency bank in Table 3.1.

Capacitance Boosting Interconnect inductance, that is usually viewed as a parasitic nuisance in LC oscillator circuits, can be used to increase the tuning range and maintain tuning monotonicity. To understand this, let us consider the reactance looking into the branch in Fig. 3.13. Writing the reactance as shown in (3.8) we note that the magnitude of X is reduced by the interconnect parasitic inductance. Therefore, the effective capacitance can now be written as shown in (3.9).

$$X = -\frac{1}{\omega RC} + \omega(2L_{int}) \tag{3.8}$$

$$C_{eff} = -\frac{1}{\omega RX} \tag{3.9}$$

The capacitance is therefore effectively magnified using the parasitic interconnect inductance. Also, because of the relative magnitudes, the parasitic capacitance when the capacitor bit is switched **off**, is hardly affected by the interconnect inductance. This can be verified by using a small value of C in (3.9). This feature can be used to increase the tuning range as well as to make the tuning characteristic monotonic. To achieve the latter, longer interconnects are used for the larger capacitor bits (Fig. 3.11 and 3.12) to compensate for the smaller switches used. The top interconnect metal is used for providing the parasitic inductance for the capacitor bank.

Note that this interconnect inductance gives rise to a higher order resonator that can have multiple modes of oscillation [84]. The resonator should be designed carefully to ensure that these higher order modes are maintained at a much higher frequency such that the startup criteria described in (3.7) fails for these parasitic oscillation modes.

Fig. 3.13 Model of the capacitor bank with inductive interconnects

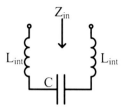

Fig. 3.14 Frequency tuning range versus capacitance in terms of the 'x', where 'x' is the capacitance of a unit switched capacitor

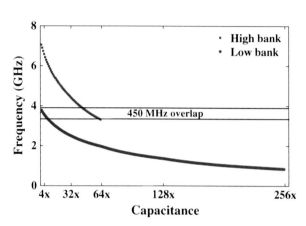

3.7 Simulation Results

The circuit was simulated using IBM's 0.13 m CMOS process in Cadence® using n-port S-parameter blocks for the interconnects and the switched-inductor. S-parameter data for these blocks were obtained from electromagnetic (EM) simulations in ADS Momentum®. Results from simulation are discussed below.

Tuning Range The frequency tuning range (FTR) obtained from simulation spans 6.21 GHz from 850 MHz to 7.06 GHz (157 %) as shown in Fig. 3.14. This is by far the largest tuning range reported in CMOS LC VCOs to date. The two frequency banks are made to overlap slightly (450 MHz) to ensure continuous frequency coverage in the face of process variations. The FTR decreases to 144 % (1 GHz to 6.2 GHz) post RC extraction of the capacitor bank.

This design may be compared to the previous version in a similar process technology which achieved 87 % tuning range. For the present design, the design methodology used for the switched-inductor and g_m cells is similar to the earlier design. However, optimization of the capacitor bank as described in this book increased the tuning range by an additional 70 %. Interestingly, apart from a slight increase in design complexity, there is no price paid for this increased tuning range.

Phase Noise The variation in phase noise over the tuning range is shown in Fig. 3.15.

Phase noise is seen to vary between −119.1 and −107.1 dBc/Hz at 1 MHz offest as shown in Fig. 3.15. As visible in the figure, when a new capacitor is switched

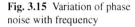

Fig. 3.15 Variation of phase noise with frequency

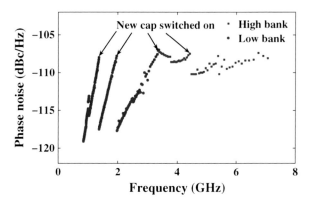

Fig. 3.16 Variation of VCO core power dissipation with frequency

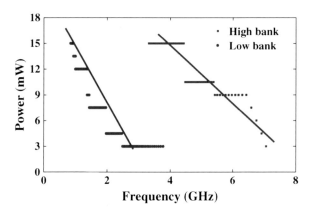

in, the phase noise remains approximately constant due to the scheme discussed in Sect. 3.6. This manual tuning accounts for a slight overhead in design as compared to conventional switched-inductor design techniques.

The peak inductor Q for this design was found to be 16 in simulation. A higher inductor Q ($\approx$ 22) can be obtained in this process by using a high resistance substrate and patterned ground shields, improving the phase noise considerably ($\approx 20log(Q_1/Q_2)^2 \approx 6dB$).

Power The variation in power is shown in Fig. 3.16 with a linear fit in each frequency bank. Similar to the technique used in the earlier design, the power dissipation is limited to a maximum of 15 mW through g_m cell sizing to improve the tuning range. The overall trends are as expected from the analysis in Sect. 3.2.

A comparison of this work with previous wide-tuning range designs is shown in Table 3.2.

Table 3.2 VCO Performance comparison

Ref	f_0 (GHz)	FTR (%)	Phase noise[a] (dBc/Hz)@1 MHz	FOM_{PFTN} (dB) [79]	Implementation μm
[79]	2.0 to 2.6	26	-125.4 to -119.4	-3.1	0.35 BiCMOS
[85]	3.1 to 5.6	58.7	-120.8 to -114.6	5.9 to 10.3	0.13 SOI
[41]	1.1 to 2.5	73	-126.5 (f_0)	5.0 to 8.5	0.18 CMOS
[86]	3.6 to 8.4	74	-104 to -101.5	-4.6 to 4.0	0.13 CMOS
[87]	2.2	92.6	-124 to -120	-2.95 to 1.05	0.13 CMOS
Design 1 (measured)	*3.3 to 8.4*	*87.2*	*-122 to -117*	*6.6 to 10.2*	*0.13 CMOS*
Design 2 (simulated)	*1 to 6.2*	*144*	*-119.1 to -107.1*	*-1.1 to 15*	*0.13 CMOS*

[a]Assuming 20 dB/decade drop with offset frequency

3.8   Conclusions

In this chapter, inductor switching was introduced as a viable solution for obtaining very wide tuning range oscillators with low phase noise. The advantages of inductor switching were identified over traditional pure capacitive tuning solutions. Significant advantages were seen to accrue from inductor switching, and a design methodology for switched-inductor oscillators was subsequently developed. Simple models were introduced to obtain further design insight into switched-inductor resonators. For a proof of concept, a new inductor switching scheme was proposed. A first prototype design based on a single switch inductor was implemented in CMOS, and was seen to simultaneously achieve a phase noise between -117.2 and -122 dBc/Hz at 1 MHz frequency offset, and an unprecedented tuning range of 87.2 % (3.3–8.5 GHz) for single inductor LC VCOs in measurement. A second prototype exploited the optimized design of a capacitor bank that increases the tuning range of switched-inductor oscillators to 157 %, which is the highest reported till date for CMOS LC VCOs. The VCO covers a frequency range spanning 850 MHz to 7.1 GHz covering the cellular, wifi and UWB bands. The increase in tuning range, due only to the optimal design of the capacitor bank, is an additional 70 %. The phase noise varies between -107.1 and -119.1 dBc/Hz at 1 MHz offset with power dissipation between 3 and 15 mW. For obtaining this performance, the capacitor switches were optimally sized, and the normally problematic parasitic interconnect inductance was used to advantage for capacitance boosting. Since these techniques do not affect the worst case phase noise or power performance, there is no apparent price paid for this increased (70 % greater than previous version) tuning range.

Chapter 4
RF Sampling and Signal Processing

4.1 Introduction

As seen in Fig. 2.12 in Chap. 2, this spectrum sensing architecture uses an RF sampler followed by discrete time signal processing in the analog domain. Specifically, passive charge domain computations are utilized for signal processing followed by digitization. For RF sampled processors, the RF sampler has historically remained a substantial bottle-neck. However, with technology scaling and subsequent improvement in switch performance, RF sampling has become a possibility in modern silicon processes. Moreover, it is possible to use charge domain sampling to leverage the inherent benefits of including an in-built anti-aliasing filter into the sampler [88], robustness to jitter [89], and the ability to vary the resulting filter notches by varying the integration period [47, 89–92]. This use of RF samplers and subsequent discrete-time processing provides a number of advantages in deep sub-micron CMOS processes [93]. Recently, other discrete time wireless radio receivers using RF sampling have been demonstrated using CMOS technology for Bluetooth [49], GSM/GPRS [48], WLAN [94], and SDR-type applications [95, 96] (Fig. 4.1).

4.2 Example Comparison

In order to appreciate the kind of power gains available in an analog signal processing architecture implemented using passive switched capacitor circuits against a digital signal processing architecture, we will consider two example architectures, and compare their power performance. Let us consider the simple and almost ubiquitous architecture shown in Fig. 4.2 (Arch A).

As shown in the figure, the incoming RF signal is first amplified, and then down-converted into real and imaginary signal paths. Each of these signal paths is low-pass filtered and digitized. The digital signal is then filtered using a complex FIR filter in the digital domain to sift out the desired signal.

B. Sadhu, R. Harjani, *Cognitive Radio Receiver Front-Ends,*
Analog Circuits and Signal Processing 115, DOI 10.1007/978-1-4614-9296-2_4,
© Springer Science+Business Media New York 2014

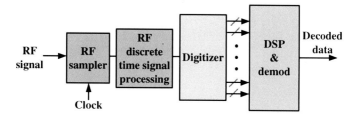

Fig. 4.1 A spectrum sensing architecture based on RF sampling and signal processing

The same function can be implemented using an alternative architecture where the complex FIR filtering is performed in the analog domain as shown in Fig. 4.3 (Arch B).

Let us now perform a comparative analysis of the power dissipation at a particular technology node. Later we will also compare the performance across technology and develop architectural insights based on the results. Also, since the architectures differ only in the last two elements, namely the complex FIR filter and the ADC, we will consider only these for our comparison. A complex N-tap FIR filtering operation can be mathematically represented through the following equation:

$$y[n] = \sum_{i=0}^{N} b_i x[n - i] \tag{4.1}$$

where the coefficients b_i and the samples x_{n-i} are complex.

Power Analysis of Arch A

1. *N-tap digital FIR filter*: For the power analysis, we will assume an M-bit ADC output (representing each $x[n]$) and C-bit coefficients (representing each b_i). The power dissipation of a complex FIR filter implemented in digital can be analyzed by breaking down the total power in terms of the power used for multiplications (P_{mul}), accumulation operations (P_{acc}) and memory operations (P_{mem}): $P_{FIR} = P_{mul} + P_{acc} + P_{mem}$. These operations can be written down in terms of the more fundamental operations of addition, shifting and memory access.

$$P_{mul} = N(M + 1)(C - 1)(P_{add} + P_{shift}) \tag{4.2}$$

$$P_{acc} = (N - 1)(M + C + \log_2 N)P_{add} \tag{4.3}$$

$$P_{mem} = (N)(M + C)P_{sram} \tag{4.4}$$

Letting E_{gate} be the energy dissipation for a minimum sized nMOS transistor gate in 65 nm technology. From an analysis of a ring oscillator in IBMs 65 nm technology, E_{gate} is found to be 1.39 nW/MHz. Using this value and assuming a static CMOS implementation, P_{tot} is calculated for $M = 12$ and $C = 6$ and plotted in Fig. 4.4 for different tap numbers (N).

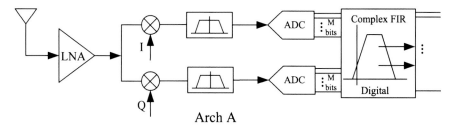

Arch A

Fig. 4.2 Receiver architecture incorporating digital domain FIR filtering

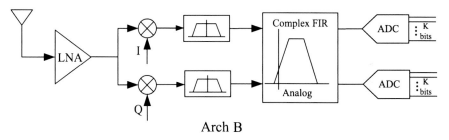

Arch B

Fig. 4.3 Receiver architecture incorporating analog domain FIR filtering

2. *M-bit ADC*: The power consumption for the ADC can be roughly calculated from the following figure of merit expression:

$$FOM = \frac{P_{diss}}{f_s 2^{ENOB}} \tag{4.5}$$

where P_{diss} is the power consumption of the ADC, f_s is the sampling frequency and ENOB is the effective number of bits. Using an empirical minimum value of 0.15 pJ/conversion for the FOM based on previous literature, the power dissipation for a 12-bit ADC with an f_s of 1 GHz is found to be 2.46 W. The total power for the FIR and ADC in Arch A with individual breakdowns for the different functionalities involved is shown in Fig. 4.4.

Power Analysis of Arch B

1. *N-tap analog FIR filter*: For implementing an N-tap analog filter, we consider the 2-tap example architecture shown in Fig. 4.5.
   The filtering operation is performed through three steps:
   a) Sampling charge onto a capacitor and scaling the sample with a coefficient b_i: This is performed using a G_m cell and series capacitors such that the voltage at the intermediate node is a scaled version of the sampled voltage. Since each sample corresponding to a time instant is used in the computation of N outputs, each time with different coefficients, in an N-tap FIR filter the charge at each time instant is sampled onto N capacitor-pairs, each with appropriate capacitive ratios. Therefore, if a capacitor combination of C_1 and C_{min} is used for charge sampling, $b_i = C_1/C_1 + C_{min}$.

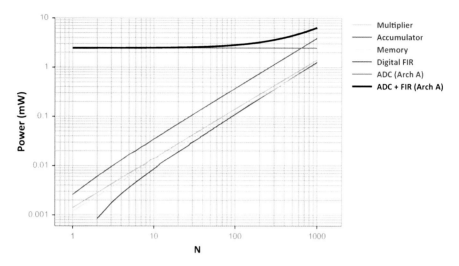

Fig. 4.4 Power dissipation for different functions in Arch A vs. number of taps in the FIR filter

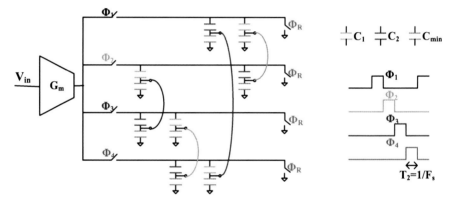

Fig. 4.5 Architecture for analog 2-tap FIR using passive switched capacitors

b) Accumulating the voltages from different samples: This can be performed by shorting together one copy of the charge sampled and scaled from each time instant. The accumulated voltage then becomes $V_{acc} = (1/N) \sum_{i=1}^{N} b_i x[n-i]$ which is a scaled version of the required result.

c) Resetting the capacitors: The capacitors need to be reset so as to be ready for the next iteration.

As a result of the two additional steps of accumulation and reset, two extra rows of capacitors are required (over and above the N rows) to accommodate a pipelined operation as shown in Fig. 4.5. The curved lines denote the accumulation operation in the appropriate clock cycle as depicted through its color. The reset switches for the intermediate nodes have not been explicitly shown to reduce clutter. For

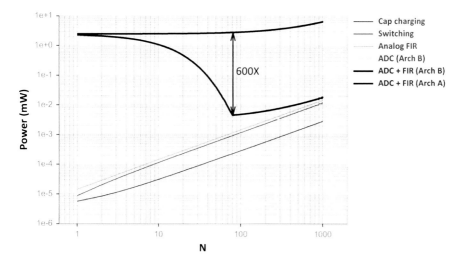

Fig. 4.6 Power dissipation for different functions in Arch B vs. number of taps in the FIR filter; also shown is the total (ADC and FIR) Arch A power dissipation for comparison purposes

a calculation of the power dissipation of the system, a constant frequency of operation (1 GHz) has been assumed (with varying N). For this, the switches had to be sized in order to maintain a constant RC time constant for all operations. The designed switch sizes allow a maximum frequency of operation of 10 GHz for the 65 nm technology used for calculations, and therefore, a 1 GHz operating frequency should be easily achievable. The architecture in Fig. 4.5 shows only real number operations. Complex number operations can be easily performed by using a few additional multiplication and scaling operations.

The power dissipation of this system can be divided into two parts:

a) Power dissipated for charging the capacitors during the sampling operation
b) Power dissipated in charging the transistor gates for all the switching operations

These values are calculated for the architecture shown, and the results are plotted in Fig. 4.6. The total power dissipated in the digital FIR block is shown in the same graph for comparison. As seen from this analysis, the power dissipation in the analog implementation of the N-tap FIR filter is more than two orders of magnitude lower than in the digital implementation. In fact, this result is hardly surprising. While, in analog, each voltage is represented and computed using a few capacitors connected in series or parallel, the same operation in digital involves an M-bit sample and C-bit coefficient and a corresponding complex algorithm to compute the result of an addition or scaling operation.

2. *K-bit ADC*: The ADC requirements in Arch B are different from those in Arch A. Since the input to the ADC is already filtered, the dynamic range requirements for the ADC (K bits, $K < M$) are reduced considerably. The blocker attenuation is assumed to be proportional to the number of taps used. Also, the bandwidth

seen by the ADCs is now reduced due to this filtering operation and a reduction factor of 1/2 has been used for calculations.

The graph below in Fig. 4.6 shows the power requirements for the FIR filter and ADC for Arch B. The total requirement for these two components of Arch A is also plotted for comparison purposes. As seen in the plot, for lower number of taps, the ADC power requirements remain high (comparable to the digital power) and therefore, does not provide a substantial benefit. However, the power requirement for Arch B reduces with N and is optimal at about 55 taps with a total power dissipation of 1 mW. Comparing this power with that in Arch A (> 2.5 W), a 600× improvement in power dissipation is achievable.

4.3   Conclusions

In this chapter, architectures based on RF sampling and high-speed charge domain processing were considered. RF sampling can be a viable technique (with feasible jitter requirements) for narrowband systems, or if charge domain samplers are employed. The fundamental concepts of sampled charge processing, and different charge domain computation techniques were introduced. An example comparison between a system relying solely on digital signal processing was compared to one that employed some filtering in the sampled charge domain. A large amount of power saving for wideband systems is expected.

Chapter 5
CRAFT: Charge Re-use Analog Fourier Transform

5.1 Introduction

Conventional software defined radios (SDR) [8] strive to digitize the RF signal and perform spectrum sensing in the digital domain. However, for wideband inputs, this translates to infeasible ADC specifications [97]. Time interleaving [70] and N-path filter-banks [73, 74] have been proposed to tackle instantaneous wideband digitization. However, while time interleaving the ADCs reduces their speed, the input dynamic range (exponentially related to ADC power) remains large. Filter-banks reduce both the speed and dynamic range (by removing out-of-band signals) of the ADCs. However, conventional filter banks are based on PLLs, mixers and low-pass filters [73–75], and can be power hungry. Moreover, harmonic mixing of signals within the SDR input bandwidth severely corrupts the channelized baseband signals [98], and signal reconstruction after digitizing the filter-bank outputs is challenging.

In this chapter, we discuss an RF sampler followed by a discrete-time Fourier transform engine to perform channelization of the wide-band RF input. The use of RF samplers and subsequent discrete-time processing, prior to digitization, provides a number of advantages in deep submicron CMOS processes including high linearity, programmability, large bandwidth, robustness to jammers, immunity to clock jitter, low power ADCs, etc. [93, 97, 99]. Recently, discrete time radio receivers using RF sampling have been demonstrated using CMOS technology for Bluetooth [49], GSM/GPRS [48], WLAN [94, 100], and SDR-type applications [95, 96].

In this technique, the discrete time DFT is used as a functionally equivalent linear phase N-path filter to perform channelization [101]. The output of each bin is effectively filtered using frequency-shifted complex *sinc* filters, followed by down-conversion. This is exactly equivalent to a mixer followed by a low pass filter scheme [101]. This scheme reduces both the speed and dynamic range of the ADCs, and, by virtue of being linear phase, allows for simple reconstruction in the digital domain using an IFFT without any loss of information. A few current-based analog DFT filters have been designed [102, 103]. However, these designs are speed-limited, and consume significant power (Table 5.3) minimizing the overall gains. Additionally, they use active devices for signal processing, and are therefore expected to be more non-linear at high operating speeds. In this work, we describe the design of a charge

B. Sadhu, R. Harjani, *Cognitive Radio Receiver Front-Ends,*
Analog Circuits and Signal Processing 115, DOI 10.1007/978-1-4614-9296-2_5,
© Springer Science+Business Media New York 2014

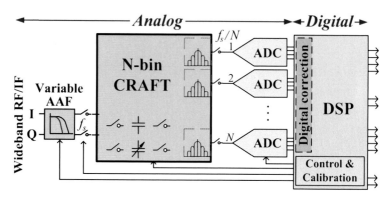

Fig. 5.1 An SDR architecture enabled by CRAFT

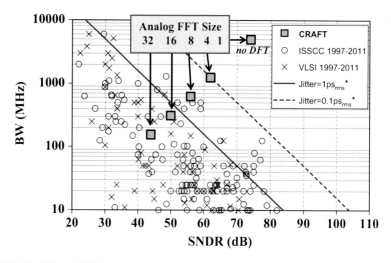

Fig. 5.2 Feasibility of ADCs vs. FFT size of the CRAFT front-end

domain DFT filter, CRAFT (Charge Re-use Analog Fourier Transform), based on passive switched capacitors. It performs an analog domain 16 point DFT running at input rates as high as 5 GS/s and uses only 3.8 mW, or 12.2 pJ per 16 point DFT conversion. By virtue of its I/Q input processing capability, it is able to transform a 5 GHz asymmetrically modulated bandwidth instantaneously. The design was first presented in [104]; this chapter includes further details of the implementation, modeling and mitigation of non-idealities and additional measurements.

Figure 5.1 shows an example architecture with a CRAFT RF front-end. We will show that this architecture reduces both the ADC speed (by N) and the required ADC dynamic range by removing out-of-band signals per ADC, at a negligible power overhead. The impact of the CRAFT front-end on the ADC input bandwidth and dynamic range for well-spread signals is shown in Fig. 5.2 where the expected ADC requirements for a 5 GS/s input are plotted amongst measured ADC implementations [61]. As seen in Fig. 5.2, the CRAFT front-end reduces the required speed of each

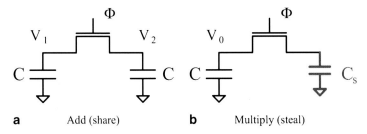

Fig. 5.3 Conceptual implementation of addition **a** and sub-unity scaling **b** operations on which CRAFT is based

ADC as well as their input dynamic range through channelization. This brings the ADC requirements from being infeasible (top right corner) toward being achievable (bottom left half) thereby solving the critical wide-band digitizing problem in SDRs. In contrast, a time interleaved ADC approach only reduces the speed requirement of each ADC without reducing the dynamic range as shown by the vertical arrow in Fig. 5.2; consequently, the total ADC power remains approximately unchanged.

5.2 CRAFT Design Concept

CRAFT operations are based on charge re-use. Once sampled, the charge on a capacitor is shared and re-shared with other charge samples such that the resulting mathematical manipulation is an in-place DFT. By basing the design only on toggling switches (transistor gates), low power and high speeds are made feasible. Additionally, the power scales with frequency, supply, and technology in a digital-like fashion.

A radix-2 FFT algorithm was used in CRAFT. The FFT computation uses only two types of operations: addition, and multiplication by twiddle factors. Note that these twiddle factors ($W = e^{-\frac{2\pi j}{16}}$) are equi-spaced points on a unit circle in the complex plane. As a result, for these scaling factors, W^k, $\Re\{W^k\} \leq 1, \forall k$ and $\Im\{W^k\} \leq 1, \forall k$. Since passive computations inherently attenuate the signal, these operations are particularly suited for sub-unity scaling.

To perform the CRAFT operations, the following charge domain computations are used. Addition is performed by sharing the charges on two capacitors as shown in Fig. 5.3a. Sub-unity multiplication is performed by stealing charge away from a capacitor using a suitably-sized stealing capacitor (C_s) as shown in Fig. 5.3b. These 2 simple operations form the basis of all operations performed in CRAFT.

The 16 point, radix-2 CRAFT operation can be represented as a linear matrix transform[1]:

$$\mathbf{X} = \mathbf{F}\,\mathbf{x} \tag{5.1}$$

---

[1] Note that any linear transform with a fixed matrix can be performed using the addition and multiplication techniques outlined above. Due to the inherent attenuation in charge domain operations, the result is a scaled version of the desired transform.

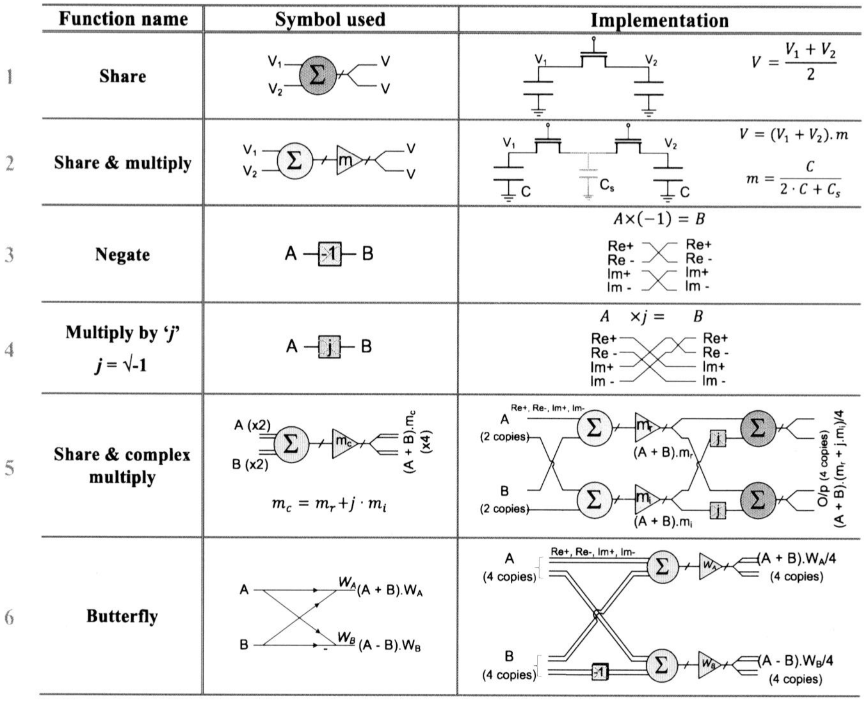

Function name	Symbol used	Implementation
1 Share	$V_1 \rightarrow \Sigma \rightarrow V$, $V_2 \rightarrow \Sigma \rightarrow V$	$V = \dfrac{V_1 + V_2}{2}$
2 Share & multiply	$V_1 \rightarrow \Sigma \rightarrow m \rightarrow V$, $V_2 \rightarrow V$	$V = (V_1 + V_2) \cdot m$, $m = \dfrac{C}{2 \cdot C + C_s}$
3 Negate	$A - \boxed{-1} - B$	$A \times (-1) = B$
4 Multiply by 'j' $j = \sqrt{-1}$	$A - \boxed{j} - B$	$A \times j = B$
5 Share & complex multiply	$A(x2) \rightarrow \Sigma \rightarrow m_c \rightarrow (A+B).m_c$ (x4), $B(x2)$, $m_c = m_r + j \cdot m_i$	
6 Butterfly	$A \rightarrow W_A (A+B).W_A$, $B \rightarrow W_B (A-B).W_B$	

Fig. 5.4 Mathematical operations in the CRAFT processor using charge-sharing

These operations are utilized in the CRAFT processor as shown in Fig. 5.4. Row 1 shows the 'Share' operation. Subsequent sub-unity twiddle-factor multiplication, if required, is performed by charge stealing as shown in row 2: 'Share and multiply' operation in Fig. 5.4. Negation is performed by swapping the positive and negative wires as shown in row 3, while multiplication by j ($= \sqrt{-1}$) is performed by swapping the appropriate wires of the real and imaginary components as shown in row 4 of Fig. 5.4. These techniques are extended to perform complex multiplication, as shown in row 5. An example butterfly operation using multiple complex multiplications is also shown in row 6. Note that each butterfly operation requires 2 clock phases. However, in the CRAFT engine, we optimize the operations such that only 5, instead of 8, clock phases are utilized for the 4 butterfly stages. This optimization is discussed further in Sect. 5.3. The optimized butterfly blocks are then used to construct the 16 point CRAFT engine shown in Fig. 5.5. As shown, the following different types of butterflies are utilized: 1-stage share (Type 1), 1-stage scalar multiply (Type 2A), 1-stage complex multiply (Type 2B), 1-stage scalar multiply (Type 3A), 2-stage complex multiply (Type 3B), and 2-stage complex multiply (Type 3C). The circuit schematics for these butterflies is shown in Fig. 5.6, and discussed in more detail in Sect. 5.3.2.

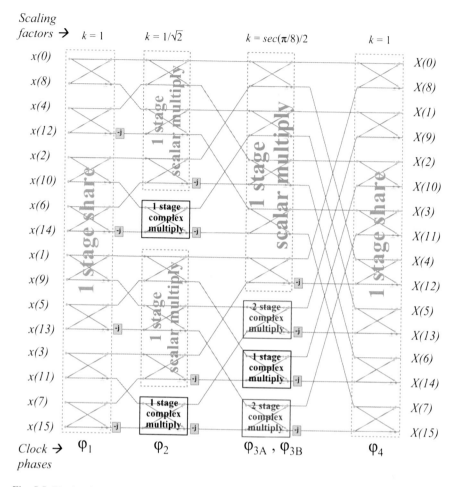

Fig. 5.5 The implemented CRAFT algorithm showing the different kinds of butterflies used and the scaling in each stage

The in-place CRAFT computations are destructive; therefore, multiple copies of each value are required for multiple operations. Since a radix-2 FFT algorithm performs the DFT with a minimum number of operations per operand (less copies required), it is selected for CRAFT.

Conceptually, the FFT is computed as follows: The input signal is sampled onto capacitors. Since each input is operated on twice in an FFT butterfly, and twice for complex operations, 4 copies of each sample are required. Also, considering I, Q (= 2) and differential (= 2) inputs, the 16 point FFT requires 16×2 (complex math) $\times 2$ (butterfly branches) $\times 2$ (I, Q) $\times 2$ (differential) = 256 sampling capacitors.

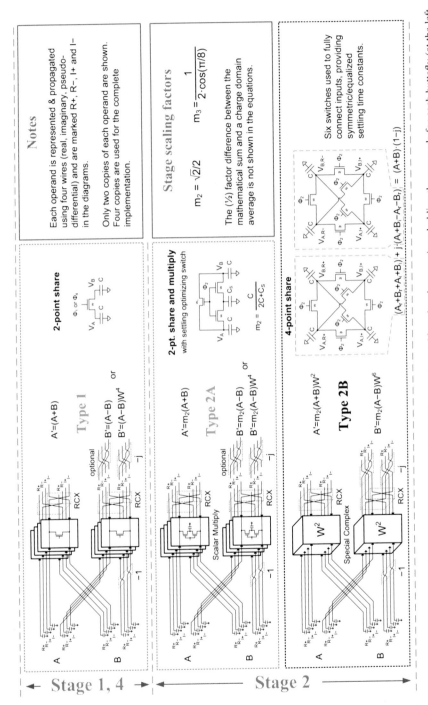

Fig. 5.6 The four FFT stages and their component types of butterflies are shown. Note that the capacitors holding the operands for each butterfly (at the left of the diagram) are the sampling capacitors

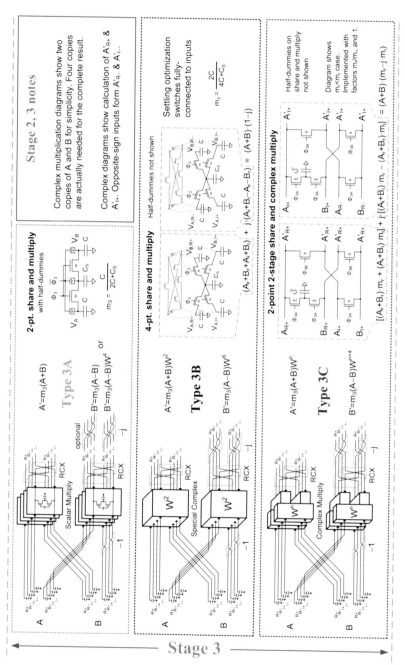

Fig. 5.7 The four FFT stages and their component types of butterflies are shown. Note that the capacitors holding the operands for each butterfly (at the left of the diagram) are the sampling capacitors

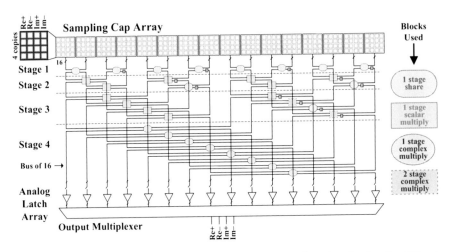

Fig. 5.8 A block-level schematic imitation of the CRAFT layout floorplan showing the different blocks used for each stage

5.3 CRAFT Implementation

The CRAFT engine is implemented as shown in Fig. 5.8. The figure closely emulates the layout implementation. The design core, shown in Fig. 5.8, includes 4 ($= \log_2(16)$) stages of FFT butterflies similar to the signal flow graph representation in Fig. 5.5. Also, in Fig. 5.8, note the different types of butterflies corresponding to those in Figs. 5.5 and 5.6. As shown, the core design is supplemented with input voltage samplers for data input, and output latches and multiplexers for data read-out. A block-level detailed description of the figure follows.

All the blocks shown are clocked and controlled by state machines. The timing diagram for the operations is shown in Fig. 5.9. The sampling operation (16 clock phases) and processing operation (4 clock phases) are allowed equal amounts of time in anticipation of an interleave-by-2 implementation. The slower processing phases allow more settling-time, lowering both the error and the power consumption.

5.3.1 Sampler

The I and Q inputs are probed into the chip and sampled on the 16 sample phases shown in Fig. 5.9. Figure 5.10c shows the pseudo-differential sampler used in CRAFT. Compared to other traditional approaches shown in Fig. 5.10a, b, this topology provides excellent signal feedthrough cancellation for differential signals.

During the sampling hold mode, the following mechanisms cause signal feed-through and leakage: capacitive feed-through due to the appreciable C_{ov}/C ratio, and sub-threshold leakage dependent on the input as well as the sampled and held voltages (see Fig. 5.10a). A traditional way to improve the isolation is shown in

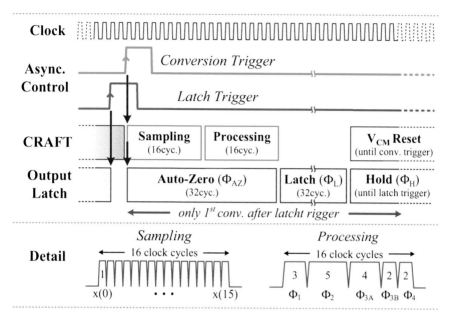

Fig. 5.9 Timing diagram showing the clock, trigger, sampling, processing, reset, and output latch clocks

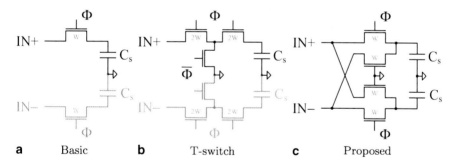

Fig. 5.10 The improved sampler (**c**), compared with 2 traditional samplers (**a**), (**c**)

Fig. 5.10b [105]. However, to maintain the same on-resistance, power consumption increases by more than 4X and the total parasitic loading of the input driver increases.

In comparison, the improved sampler configuration (Fig. 5.10c) uses matched cross-coupled feedthrough paths from the differential inputs to cancel the capacitive signal feed-through at the output node without any additional switching or active power consumption. Moreover, sub-threshold input leakage effects are differentially-suppressed for small hold voltages. Simulations suggest that the improved design in Fig. 5.10c cancels feed-through and leakage and provides isolation primarily limited by local device and layout mismatch. For example, a matching of 0.1% (3σ) achieves 75 dB isolation compared to Fig. 5.10a.

As shown in Fig. 5.8, each input sample is stored using a set of 16 capacitors (4 copies of pseudo-differential, complex inputs). For a 16 point FFT, 16 such sets (total of 256 capacitors) are used. The 256 sampling switches are designed to run at 5 GS/s with a total differential input swing of 1.2V and a 60 dB SFDR in simulation.

The design samples and processes signals using a pseudo-differential representation; two capacitors, each with voltages ranging from 0 to $2V_{cm} = V_{cm} \pm V_p$, give a signal range of $V_{pp,diff} = 4V_{cm}$. Advantages of this arrangement include: maximized swing while using nMOS-only switches; issues inherent to below-ground swing are avoided; wire swapping can be used for performing negation without bottom-plate switches; capacitors can use a shared bottom-plate for a savings; and signal-independent non-idealities, like clock-feedthrough, appear as common-mode. Although not used in this work, it is also possible to use bottom plate sampling to alleviate charge injection errors in the passive sampler [106].

The choice of the size of each sampling capacitor is a trade-off between the speed and the sampled kT/C noise in the circuit. For capacitor size selection, a -60 dB floor goal is selected. The sampling capacitors are sized at $200fF$ each such that the total calculated output-referred kT/C noise contribution by the CRAFT engine due to the reset, sampling, and processing operations is 63.3dB below the output full-scale. Details of the noise contribution of the processing operations is discussed in Sect. 5.4.

As shown in Fig. 5.8, 16 buses, each 16 wide, are connected to the sampling capacitors, and run through the CRAFT core. These wires are always connected to the sampling capacitors; consequently, their parasitic capacitance forms a part of the sampler and needs to be accurately matched. Additionally, since the operations are performed in-place, the outputs appear on the sampling capacitors at the end of the processing phase.

5.3.2   CRAFT Core

The CRAFT core performs the FFT operation in 4 ($= \log_2 (16)$) stages as represented in Fig. 5.5. The processing clock phases are shown in Fig. 5.9. As described in Fig. 5.4, each butterfly is conceptually implemented using switches and scaling capacitors that perform a set of share and multiply operations on the input samples. Moreover, the *a priori* knowledge of the exact FFT operations are exploited to hard-wire a number of modifications that reduce the number of clock phases required per stage, and limit the signal attenuation inherent to passive computation. The implementation details of these improvised butterflies in each of the 4 stages is shown in Fig. 5.6 and described below in detail Fig. 5.7.

Stage 1 The first stage comprises only share operations as shown in Fig. 5.5. The butterflies in this stage are almost identical, as shown in Figs. 5.5 and 5.8 with the implementation in Fig. 5.6 (Type 1). As shown in Fig. 5.6 (Type 1 butterfly), corresponding ($\Re\pm$ & $\Im\pm$) copies of each operand A and B are shared during stage 1 processing. Since the magnitude of the multiplicand in this stage is unity, the multiplication stage can be eliminated. RC error cancellation, as discussed in Sect. 5.4.2, is performed using wire-swapping. This stage consumes a single processing-clock cycle as shown in Fig. 5.9.

Stage 2 The second stage comprises two types of operations: share and multiply by unity, and by $W^2 = \frac{1}{\sqrt{2}} - \frac{1}{\sqrt{2}}i$, as shown in Fig. 5.5. We note that a share followed by multiplication with a twiddle factor $(m_r - jm_i)$ with equal real and imaginary parts $(m_r = m_i = m)$ can be simplified as $[(A_r + jA_i) + (B_r + jB_i)] \cdot m(1 - j) = (A_r + B_r + A_i + B_i) \cdot m + j(-A_r - B_r + A_i + B_i) \cdot m$ [107]. As a result, the 2-stage complex multiplication can be replaced to a single-stage 4-operand share and multiply operation. This can be further simplified to a single symmetric share operation to reduce the attenuation and equalize the settling, as shown in Fig. 5.6 (Type 2B). The mathematical scaling due to a 4-share operation is $\frac{1}{4}$. Scaling this to the intended attenuation of $\frac{1}{\sqrt{2}}$, we note that a unity gain needs to be replaced by a share and multiply operation that scales the result by $\frac{1}{2\sqrt{2}}$ as shown in Fig. 5.6 (Type 2A). The third switch, in Type 2A, is used for improving the differential settling as discussed in Sect. 5.4.2. The innovation using a 4-share operation eliminates one clock phase, thereby providing for more settling-time, and minimizes the mathematical attenuation inherent to charge-computations by $\sqrt{2}$.

Stage 3 The third stage comprises share and multiply operations with twiddle factors $W^1 = j \cdot W^5 = \cos\left(\frac{\pi}{8}\right) - j \cdot \sin\left(\frac{\pi}{8}\right)$, and $W^3 = j \cdot W^7 = \sin\left(\frac{\pi}{8}\right) - j \cdot \cos\left(\frac{\pi}{8}\right)$, as shown in Fig. 5.5. These butterflies require the complete complex multiplication stage shown in Fig. 5.4 and consume 2 clock cycles: ϕ_{3A} and ϕ_{3B} in Fig. 5.9. Also, for the 4 twiddle factors, only two types of stealing capacitors, in conjunction with 'j' wire-swaps, are used to reduce mismatch. The implementation of these butterflies is shown in Fig. 5.6 (Type 3C). The complex multiplication is implemented in 2 stages. The entire operation is scaled by $1/m_i$ reducing the attenuation by $\sec(\pi/8)$. Consequently, only one scaling capacitor is necessary for this butterfly type, as shown.

The share and multiply by unity is performed using a share and multiply operation similar to stage 2 to equalize the total stage scaling as shown in Fig. 5.6 (Type 3A), and utilize both ϕ_{3A} and ϕ_{3B} for their operations. Half-dummy switches are used as shown to cancel clock feed-through and reduce charge injection.

Multiplication by W^2 is performed using a 4-point share and multiply (to adjust for the total stage scaling) similar to stage 2. The optimized switching configuration is determined and shown in Fig. 5.6 (Type 3B). Extra switches are added for improving the differential settling. Half-dummy switches added for clock feed-through cancellation are omitted to avoid clutter in the diagram.

Stage 4 The fourth stage comprises only share operations as shown in Fig. 5.5. These butterflies are identical to the butterflies in stage 1 and are not redrawn in Fig. 5.6.

Using the improvised share and multiply techniques, only 5 processing phases are used. Moreover, the total attenuation is limited to $0.38 (= 1 \times \frac{1}{\sqrt{2}} \times \frac{1}{2}\sec\left(\frac{\pi}{8}\right) \times 1)$ as shown in Fig. 5.5. This attenuation is 6.1X (15.7 dB) superior compared to an unmodified implementation.

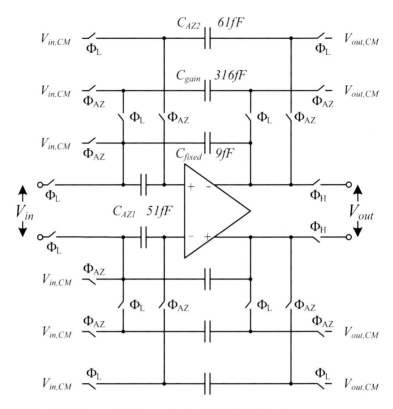

Fig. 5.11 One of the 32 output latches used to latch the CRAFT results

5.3.3 Output Latch

On the far end of the core, the wires connect through switches to the operational transconductance amplifier (OTA) based analog latches that save the outputs prior to being read out (Fig. 5.8). *These analog latches are needed only for measurement purposes to prevent capacitor leakage prior to the slow sequential read-out and to drive the large load presented by the off-chip ADCs.* Once the ADCs are integrated, the CRAFT engine will directly interface with the ADCs, obviating the need for these latches. To match the CRAFT performance, a two-stage, folded-cascode, differential OTA with 70 dB gain and 900 MHz UGB is designed. The OTA is used in the differential switched capacitor analog latch shown in Fig. 5.11 with a closed loop gain of 2.5 providing a − 7.8 dBV output. The corresponding clock phases used are shown in Fig. 5.9. The latch performs offset cancellation and switched-capacitor based common-mode feedback during the CRAFT sampling and processing phases (ϕ_{AZ}), and latches the CRAFT output with a 10τ settling accuracy during the next 32 clock phases (ϕ_L). The output is then held (ϕ_H) until the external measurement system reads the outputs. Thirty-two (16×2 for real and imaginary) of these latches capture the FFT output Fig. 5.12.

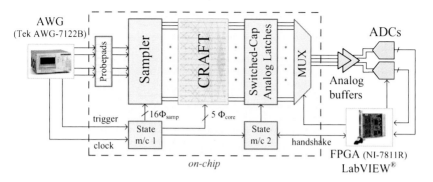

Fig. 5.12 Test setup for the CRAFT processor (on and off chip)

5.4 Circuit Non-Idealities and Mitigation

As expected, the design relies heavily on digital state machines. Also the design's regularity and complexity approach that of digital designs. However, despite these similarities, the CRAFT engine remains vulnerable to analog circuit non-idealities including noise, matching, and non-linearity. Consequently, accurate modeling of non-idealities is critical. Switched capacitor circuit noise, charge-injection, and charge-accumulation were not adequately modeled in the Spectre models available. Moreover, in order to isolate the impact of individual non-idealities per stage, it was necessary to enable/disable them independently in simulation. For this design, CRAFT-specific models of dominant non-idealities were developed in MATLAB®. Each non-ideality is modeled as an independent error source that can be enabled in simulation to isolate the impact of each error source.

The different operations in CRAFT can be divided into the initial sampling phase, and the subsequent processing phases. In this section, the effect of noise in the sampler as well as the core is discussed. This is followed by a discussion on incomplete settling due to the high speed of operation. Other non-linearities such as clock feed-through and charge injection are also modeled and mitigated [108] using techniques outlined in [61, 109, 110] but have been omitted from this discussion in the interest of space.

5.4.1 Noise

Sampling Noise

Motivation During a voltage sampling operation, noise presents itself as a final-value disturbance with a power kT/C. In this design, 4×2 capacitors, $200\,fF$ each, are used to sample 4 copies of each input pseudo-differentially. The full-scale input is $V_{pp,\text{diff}} = 1.2V$. This sampler yields a sampled-noise voltage of $\sqrt{V_{n,rms}^2} = V_{n,rms} = 144\,\mu V$ on each of the 4 single-ended copies (-63.4 dBFS).

As the noise is uncorrelated, averaging these copies at the output and forming a differential output gives a total noise of $V_{n,rms} = 144 \cdot \frac{1}{\sqrt{4}} \cdot \sqrt{2} = 102\mu V$ (-72.4 dBc) for a full-scale 1-tone input.

Modeling Sampling noise effects are included in our MATLAB® system simulation model. After sampling, a Gaussian random variable with $\sigma = V_{n,rms} = \sqrt{kT/C}$ is added to each capacitor's final value.

Mitigation The sampling capacitor size sets both the sampler's noise as well as the baseline for the processing noise because the same capacitors are used for computations. The capacitor size is selected to be $200fF$ based on the simulated total output-referred noise.

Processing Noise

Motivation Similar to the sampling operation, noise from the CRAFT core switches corrupts the computations. At the end of every share operation, kT/C noise power is added to each output copy. Interestingly, this instantaneous sampled noise on the two capacitors arising from a single switch is completely correlated (equal magnitude, opposite signs). Similarly, the noise sampled on the capacitors during a share and multiply operation arises from the same switches, and are therefore partially correlated. Also, any noise in the input operands (e.g., from the previous stage of operations) are averaged during an operation.

Modeling For modeling, Gaussian random variables are used that are distributed in magnitude and sign on the output copies in exactly the manner the noise-transfer functions dictate for both reset and processing operations. This generates the proper noise correlation that, averaged over multiple simulations, provides the expected output-referred noise power.

Mitigation Noise in the later stages of the CRAFT engine is reduced due to copy averaging among four copies of each output to reduce the noise power by 4. Moreover, correlation ρ among copies is exploited for noise reduction by averaging before the latching operation.

Total Output Noise

The noise contribution of each stage is computed analytically and tabulated in Table 5.1. The attenuation ($A_{v,out}$) reduces the output-referred noise by $A_{v,out}^2$. The single-ended, copy-averaged noise, including the residual noise from the stealing capacitor reset operation, is computed as shown in the last column yielding a total output-referred noise of $0.26 \cdot \frac{kT}{C}$ (-63.9 dBFS per differential $\Re$, $\Im$ output for a noise EVM of -61.9 dBFS).

Table 5.1 Summary of noise contribution in CRAFT

Stage	Noise sources	$\overline{V_n^2}$ per copy	$A_{v,out}$	ρ_{out}	$P_{n,out} = \left(\frac{1}{4}\right) \cdot \overline{V_n^2} \cdot A_{v,out}^2 \cdot (1 + \rho_{out})$
	Sampler (4-copy)	$1.000 \cdot \frac{kT}{C}$	0.38	0	$0.0366 \cdot \frac{kT}{C}$
1	2pt. share	$0.500 \cdot \frac{kT}{C}$	0.38	0	$0.0183 \cdot \frac{kT}{C}$
2	2pt. sh./scale $(m = 0.707)$	$0.750 \cdot \frac{kT}{C}$	0.54	0	$0.0549 \cdot \frac{kT}{C}$
3	2pt. sh./scale $(m = 0.541)$	$0.854 \cdot \frac{kT}{C}$	1	-0.29	$0.1521 \cdot \frac{kT}{C}$
4	2pt. share	$0.500 \cdot \frac{kT}{C}$	1	-1	0
	Total output noise				$0.2619 \cdot \frac{kT}{C}$

5.4.2   Incomplete Settling

Motivation For an RF discrete time signal processor, very high speeds are mandated. Increasing the switch size to allow better settling not only increases the power consumption but also causes larger charge-injection and clock-feedthrough errors. For example, in CRAFT, stages 1–4 have average simulated settling-time to time-constant ratios of $\Gamma_{S1} = 7, \Gamma_{S2} = 4, \Gamma_{S3} = 5$, and $\Gamma_{S4} = 4$ respectively [111]. For the modeling and mitigation sections, first, a single-ended settling scenario is considered. This is then extended to the pseudo-differential operation in CRAFT.

Modeling (i) Single-ended settling: A 2-point share and multiply operation, as shown in Fig. 5.4, has the settling response of equation (5.2). The input capacitors (C), with initial voltages v_{a0} and v_{b0}, are connected to the stealing capacitor $(C_s,$ with no initial voltage: $v_{s0} = 0)$ by switches of resistance R_{sw}, where the scaling factor, $m = \frac{2}{(2+C_s/C)}$. Therefore,

$$
V_a(t) = \overbrace{\frac{1}{2}(v_{a0} + v_{b0})\,m\left[1 + \left(\frac{1}{m} - 1\right)e^{-t/\tau_s}\right]}^{\text{scaled-sum term}} + \overbrace{\frac{1}{2}(v_{a0} - v_{b0})\,e^{-t/\tau_d}}^{\text{difference term}} \tag{5.2}
$$

$$
V_b(t) = \underbrace{\frac{1}{2}(v_{a0} + v_{b0})}_{\text{ideal result}}\,m\underbrace{\left[1 + \left(\frac{1}{m} - 1\right)e^{-t/\tau_s}\right]}_{\text{scaling error factor}} - \underbrace{\frac{1}{2}(v_{a0} - v_{b0})\,e^{-t/\tau_d}}_{\text{difference settling error}}
$$

The difference and sum-settling time-constants, τ_d and τ_s respectively, are

$$
\tau_d = R_{sw}C \qquad\qquad\qquad \tau_s = R_{sw} \cdot (C \parallel C_s/2) = R_{sw}C \cdot (1 - m)
$$

(ii) Pseudo-differential settling: Pseudo-differential operands are described by the relationships below, where V^+ and V^- represent the positive and negative components of the operand and are stored on separate capacitors.

$$
V_A(t) = V_a^+(t) - V_a^-(t) \qquad\qquad V_B(t) = V_b^+(t) - V_b^-(t) \tag{5.3}
$$

The time-domain settling response of these pseudo-differential operands is shown in Fig. 5.13 for $m = 1$ (sharing operation). In general, for a share and multiply operation, using $m' = m[1 + (\frac{1}{m} - 1)e^{-t/\tau_s}]$ and $\epsilon = \frac{1}{2}(v_{a0} - v_{b0})e^{-t/\tau_d}$, we can write,

$$V_A(t) = \overbrace{\left[\frac{1}{2}\left(v_{a0}^+ + v_{b0}^+\right)m'^+ + \epsilon^+\right]}^{V_a^+(t)} - \overbrace{\left[\frac{1}{2}\left(v_{a0}^- + v_{b0}^-\right)m'^- + \epsilon^-\right]}^{V_a^-(t)} \tag{5.4}$$

$$= \left(v_{a0}^+ + v_{b0}^+\right)\underbrace{\frac{1}{2}(m'^+ + m'^-)}_{m'_d} + \underbrace{(\epsilon^+ - \epsilon^-)}_{\epsilon_d} = \left(v_{a0}^+ + v_{b0}^+\right)m'_d + \epsilon_d$$

where $m'_d = m[1 + (\frac{1}{m} - 1)\frac{1}{2}(e^{-t/\tau_s^+} + e^{-t/\tau_s^-})]$, and $\epsilon_d = \left(v_{a0}^+ - v_{b0}^+\right)\frac{1}{2}(e^{-t/\tau_d^+} + e^{-t/\tau_d^-})$.

Mitigation (i) Single-ended settling: The settling accuracy can be improved using a third switch ($R_{sw,3}$) for the share and multiply operation, as shown in Fig. 5.6 (Type 2A). This improves the differential settling error by providing an alternate settling path. Note, from Fig. 5.6 (Type 2A), the widths of the main sharing switches (having a resistance, R_{sw}) effectively occur in series for the differential settling equation. As a result, increasing the third switch width (having a resistance, $R_{sw,3}$) improves the differential settling with twice the power efficiency as compared to increasing the main switch width: $\tau'_d = (2R_{sw} \parallel R_{sw,3}) \cdot (C/2)$.

(ii) Pseudo-differential settling: Considering the pseudo-differential settling expressions under small operand swings, a method for RC settling error cancellation (RCX) is developed. Recognizing that V_a^- and V_b^- settle to the same value but have ϵ error of opposite sign, they can be interchanged to form the new differential operands below (compare with equation 5.3),

$$V_{A,RCX}(t) = V_a^+(t) - V_b^-(t) \qquad\qquad V_{B,RCX}(t) = V_b^+(t) - V_a^-(t)$$

This is implemented using only wire swapping as shown in Fig. 5.13 greatly reducing differential settling error. While the psuedo-differential "inputs" ($+, -$) to the operation are (V_a^+, V_a^-) and (V_b^+, V_b^-), the capacitors holding the "output" results (2 copies) are (V_a^+, V_b^-) and (V_b^+, V_a^-). This changes the ($e^{-t/\tau_d^+} + e^{-t/\tau_d^-}$) term in ϵ_d into ($e^{-t/\tau_d^+} - e^{-t/\tau_d^-}$), allowing the differential-settling error to be canceled when $\tau_d^+ \approx \tau_d^-$. Also,

$$\epsilon_{d,RCX} = (\epsilon^+ + \epsilon^-) = (\epsilon^+ - \epsilon^-)\left[\frac{\epsilon^+ + \epsilon^-}{\epsilon^+ - \epsilon^-}\right] = \epsilon_d \cdot \tanh\left[t\left(\frac{\tau_d^+ - \tau_d^-}{2\tau_d^+ \tau_d^-}\right)\right] < \epsilon_d$$

$$= (\epsilon^+ - \epsilon^-)$$

shows that RCX always yields a net improvement in inter-copy error, and is used in all the CRAFT butterflies as seen in Fig. 5.6. Simulations show a net improvement of about 10dB in the FFT settling error. It is important to understand that the inter-copy error is not actually being cancelled, it is just partially translated into a common-mode component. The common-mode rejection of the stages that follow reject this settling error.

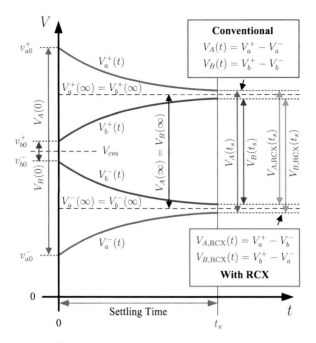

Fig. 5.13 Reduction in differential settling error using RCX ($m = 1$)

Note that the switch resistance is voltage dependent causing a spread in the re-alized time-constants. Consequently, effective time-constants based on simulations are calculated and utilized in the incomplete-settling and RCX equations to ensure our computation accuracy requirements.

5.5 Measurement Results

The design has been implemented in the IBM 65 nm CMOS process. Measurement results are shown in Figs. 5.14, 5.15 and 5.16a. The measurements shown are after calibration to compensate for systematic offsets due to parasitics. The calibration process used is detailed below.

5.5.1 Calibration

The accuracy of the CRAFT operations is dependent on the matching between the capacitors used to realize it. Any systematic mismatch between the capacitors causes computation errors that reduce the dynamic range of the system. However, since these errors are systematic, and input independent, it is possible to calibrate for them in the digital domain. For measurements a simple calibration technique can be used and is described below.

Fig. 5.14 CRAFT outputs
with a 362.5 MHz 1-tone
input sampled at 5 GS/s

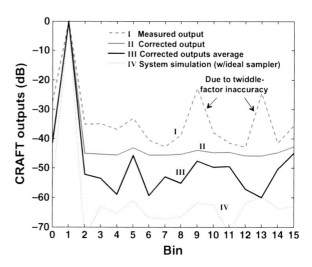

First, the non-idealities in the CRAFT operation are represented using a modified
FFT matrix, $\mathbf{F}'$, that includes the effects of mismatch and represents the non-ideal
CRAFT operation. The resultant outputs are represented by $\mathbf{X}':\mathbf{X}' = \frac{1}{k} \mathbf{F}' \mathbf{x}$. A
calibration scheme is constructed by observing that the FFT matrix $\mathbf{F}$ comprises 256
elements. Consequently, a set of 16 mutually independent inputs, vectors $\mathbf{x_i}$ with
entries $x_i(t)$, constitute 256 independent equations. Using the measured results $\mathbf{X}'_i$,
with entries $X'_i(n)$, the non-ideal $\mathbf{F}'$ matrix can be determined.

For convenience, we generate 16 orthogonal inputs comprising 16 separate 1-tone
signals, each centered on a bin i. Note that in an on-chip implementation, perfect
tones are not easily available. However, preloaded samples of linearly independent
(not necessarily orthogonal) inputs that require a low resolution DAC can be used
instead to provide similar calibration accuracy.

After determining an estimate of $\mathbf{F}'$, a one time correction matrix $\hat{\mathbf{H}}$ is computed,

$$\hat{\mathbf{F}}' = \mathbf{X}'\hat{\mathbf{x}}^{-1} \quad \Rightarrow \hat{\mathbf{H}} = \mathbf{F}(\hat{\mathbf{F}}')^{-1} = \mathbf{F}\hat{\mathbf{x}}(\mathbf{X}')^{-1}$$

where $\hat{\mathbf{x}}$ is the calibration input and $\mathbf{X}'$ is the calibration output. All subsequent
measurements across different magnitude and frequency inputs are then corrected
by applying $\hat{\mathbf{H}}$ to $\mathbf{X}' \Rightarrow \hat{\mathbf{X}} = \hat{\mathbf{F}}x = \hat{\mathbf{H}}\mathbf{F}'x = \hat{\mathbf{H}}\mathbf{X}'$.

5.5.2   Test Setup

The test setup is shown in Fig. 5.12. I and Q inputs are generated using the Tektronix
AWG-7122B arbitrary waveform generator and input to the CRAFT engine using
50Ω terminated probe-pads, which feed the sampler array. The latched outputs are
externally buffered, and captured by external ADCs controlled by an FPGA (NI-
7811R) programmed using LabVIEW®.

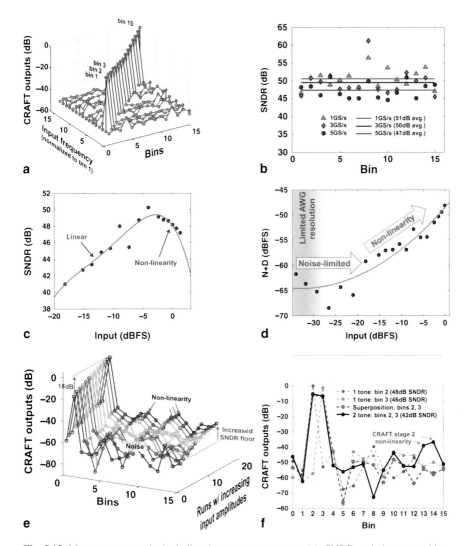

Fig. 5.15 Measurement results including 1-tone measurements: (**a**), SNDR variation across bins, and frequencies: (**b**), SNDR, and noise and distortion across input amplitudes: (**c**), (**d**), (**e**), and results of a 2-tone test: (**f**)

Note that the CRAFT processor runs at an input/output rate of 5 GS/s per *I* and *Q* channel. Additionally, the outputs are analog values. Due to the very high speed of operation, and the limitations in the number of I/O pins, the outputs are first latched using the OTA-based analog latches described in Sect. 5.3.3, and multiplexed out at a slower rate limited to 40 MS/s. The CRAFT speed is not compromised due to the output read-out limitation. This output multiplexing and read-out is controlled by an FPGA (NI-7811R) programmed using LabVIEW®. The outputs are first buffered,

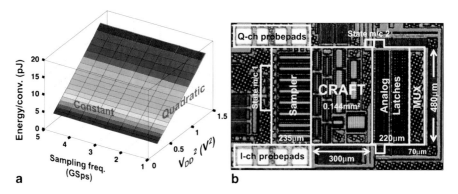

Fig. 5.16 The energy consumption relation with frequency and supply: (**a**), and a die photograph: (**b**)

then digitized by external ADCs, and finally captured by the FPGA. Using this scheme, RF inputs that are synchronous as well as asynchronous to the clock can be captured and phase aligned. The latter is particularly important to emulate practical scenarios.

5.5.3 On-Bin 1-Tone Measurements

Figure 5.14 shows the CRAFT output magnitudes ($\sqrt{\Re^2 + \Im^2}$) with a 312.5 MHz ($= \frac{5\,\text{GHz}}{16}$) 1-tone input at 5 GS/s. Curve I shows the measured uncalibrated output magnitude across 16 bins. Curve II in Fig. 5.14 shows the calibrated plot depicting the circuit noise floor (including the integrated noise of the analog latches) at -46 dB. To explore the non-linearity floor, a synchronous average over 500 measurements is used. The resultant Curve III shows the non-linearity floor with 43 dB SFDR[2] and 48 dB SNDR[3]. Note that the SNDR and SFDR also signify the out-of-band rejection of the FFT as a filter. The non-linearity predicted by simulations of the standalone CRAFT engine is shown in Curve IV. Note that Curve III not only includes the non-idealities in CRAFT, but, unlike Curve IV, also includes the non-idealities of the

---

[2] SFDR for a 1-tone test is calculated as the difference between a full-scale on-bin signal and the largest off-bin output.

[3]

$$SNDR = 20 \times \log_{10}\left(\frac{\sqrt{\sum_{k=1}^{N} V_{ideal}^2(k)}}{\sqrt{\frac{1}{N}\sum_{k=1}^{N}\{V_{meas}(k) - V_{ideal}(k)\}^2}}\right)$$

where N is the number of FFT bins. Since the outputs will be observed/digitized on a per bin basis, the total (noise + distortion) in the denominator is averaged over the N bins to yield the average (noise + distortion) per bin.

Table 5.2 Table of CRAFT
SNDR and SFDR with
single-tone, on-bin, 0 dBFS
inputs

(dB)		Min.	Max.	Mean	Sigma
1 GS/s	*SNDR*	46.6	56.5	*50.6*	2.7
	SFDR	37.7	47.8	*43.2*	4.1
3 GS/s	*SNDR*	56.5	61.2	*49.5*	3.8
	SFDR	35.8	45.5	*42.1*	5.0
5 GS/s	*SNDR*	44.5	51.1	*47.3*	2.3
	SFDR	35.6	43.3	*40.5*	2.3

8-bit resolution AWG inputs, the sampler, and systematic and random sampling jitter. Unfortunately, despite the use of state-of-the-art test equipment, the limitations in the input and output test setup severely limit the observable non-idealities in CRAFT.

SNDR Variation

One tone measurements for all bins were performed as shown in Fig. 5.15a, and the resulting SNDR at 1, 3, and 5 GS/s for a 1-tone input frequency placed at different bins is shown in Fig. 5.15b. The average SNDR across bins is $\approx$ 50 dB at 1 & 3 GHz and degrades to 47 dB at 5 GHz; SNDR better than 45 dB is maintained across all frequencies. The SNDR and SFDR measurements are tabulated in Table 5.2. This provides 7–8 bits of spectrum sensing resolution over a 5 GHz (2.5 GHz×2 due to I, Q inputs) frequency range in the digital back-end.

Figure 5.15c plots the output SNDR versus the input amplitude. A fourth-order fit shows a linear SNDR improvement with increasing amplitude before being limited by a combination of the circuit non-linearities, and the AWG resolution (8 bits).

The CRAFT output magnitudes with varying input amplitudes for a 1-tone, on-bin 312.5 MHz input at 5 GS/s are shown in Fig. 5.15e. The input is varied over an 18 dB range. As seen in Fig. 5.15d, the circuit is limited by the noise floor for low input amplitudes, while it becomes non-linearity limited at larger input amplitudes. Note that Curve III in Fig. 5.14 represents a cross-section of Fig. 5.15e.

5.5.4 On-Bin 2-Tone Measurements

Results from a 2-tone test with tones on adjacent bins are shown in Fig. 5.15f. Assuming a preceding AGC, the maximum (time-domain) amplitude of the 2-tone signal is normalized to that of a single tone in the 1-tone test as shown. Two 2-tone measurements and their superposition (normalized for same maximum amplitude) are also shown. The difference between the superposition and the measured 2-tone output is indicative of the additional non-linearity introduced. The relative increase in bins 13 and 14 is likely to be due to uncorrected twiddle factor errors in stage 2 of the CRAFT engine. Note that the apparent effect of these twiddle factor errors, leaking signal on to the negative of the signal frequency, is identical to the effect of I-Q mismatch errors in a homodyne receiver.

Table 5.3 Table for comparison with other FFT implementations

Ref	Domain	bins	SNDR (dB)	Power (mW)	Speed (GS/s)	E/conv. (pJ/conv.)[a]	E/conv. ratio[a]
[102]	Current	64	–	389	1.2	345.8	28X
[103]	Current	8	36	19	1.0	405.3	33X
[112]	Digital	128	51[†]	175	1.0	1600	131X
This work	*Charge*	*16*	*47[‡]*	*3.8*	*5.0*	*12.2*	*1X*

[a] Scaled for complexity similar to the scaling used in [103]
[†] 8.5 bit ENOB assumed for 10 bit internal word length
[‡] After twiddle factor correction

5.5.5 Power Consumption

The CRAFT core consumes 12.2 pJ/conv. and uses 3.8 mW of power when interleaved by 2 for a 5 GS/s input and 5 GS/s output. Measurements of the energy consumption versus supply voltage and frequency are shown in Fig. 5.16a. These measurements clearly show the expected digital-like relationship of the CRAFT energy with frequency and supply voltage. This further corroborates our premise that CRAFT is expected to respond favorably to technology scaling.

Figure 5.16b shows a die photograph of the CRAFT chip. The core occupies an area of 300 μm × 480 μm = 0.144 mm^2 as shown.

5.6 Conclusion

This work describes a wideband ultra-low power RF front-end channelizer based on a 16 point FFT. The design is based on a charge re-use technique that enables it to run at 5 GS/s with a 47 dB SNDR capable of transforming a 5 GHz signal (I/Q) while consuming only 3.8 mW (12.2 pJ/conv.).

The current chapter details the design of the CRAFT engine. It also describes the interface circuitry, and the test setup required to test such a high-speed, high dynamic range system. Non-idealities in the CRAFT computations are discussed, analytical models derived, and new circuit techniques developed for mitigating these. These techniques can be easily extended to other passive switched capacitor circuits to improve their performance. Measurement results are then presented.

Table 5.3 compares the CRAFT performance with one digital and two analog domain FFT implementations. As shown, CRAFT operates at speeds 5X faster than previous state-of-the-art designs. Additionally, it consumes better than 28X lower energy. As an RF channelizer, it is expected to reduce digitization requirements enabling wide-band digital spectrum sensing. As a result, it advances the state-of-the-art for wide-band SDR architectures.

Chapter 6
Conclusions

The realization of the alluring vision of a cognitive radio requires novel and disruptive RF circuit architectures, and innovative circuits. This has motivated a deluge of exciting research in this area in the past decade, and will continue to spur further research into the next. At this time, a highly reconfigurable RF architecture seems feasible by employing new wideband circuits that have recently been developed. However, the realization of the ideal cognitive radio, with the level of versatility described by early descriptions [7], requires further innovation, probably spanning the next decade.

In this book, a number of circuit architectures and techniques in recent literature have been discussed. Their suitability for cognitive radio applications has been briefly reviewed. New architectures and circuits have been described, and results from early prototypes have been presented for demonstration.

6.1 Remarks

Based on the current research directions, the following observations regarding on-chip SDR architectures for portable cognitive radios can be distilled. These remarks are summarized in Figs. 6.1 and 6.2. The aspects described in this book are highlighted in bold font.

SDR Signaling

1. For flexibility through reconfigurability in the frequency domain, frequency agility is an important characteristic. Wide-tuning, frequency agile architectures and circuits are critical for signaling applications.
2. To avoid the reciprocal mixing of large blockers onto small signals in wideband architectures, low LO phase noise is important.
3. Programmable RF filters are required in the front-end to provide the required frequency agility while relaxing the linearity requirements of the circuits that follow.
4. Programmable baseband filters are required to handle multiple radio-access technologies.

B. Sadhu, R. Harjani, *Cognitive Radio Receiver Front-Ends,*
Analog Circuits and Signal Processing 115, DOI 10.1007/978-1-4614-9296-2_6,
© Springer Science+Business Media New York 2014

SDR signaling

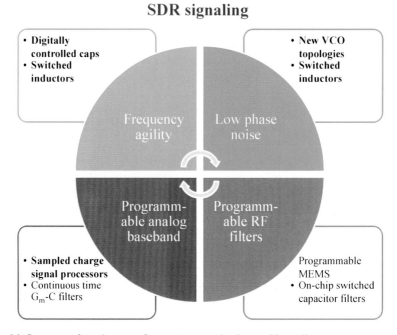

- Digitally
  controlled caps
- Switched
  inductors

- New VCO
  topologies
- Switched
  inductors

Frequency
agility

Low phase
noise

Programm-
able analog
baseband

Programm-
able RF
filters

- **Sampled charge
  signal processors**
- Continuous time
  G_m-C filters

Programmable
MEMS
- On-chip switched
  capacitor filters

Fig. 6.1 Summary of requirements for spectrum sensing in cognitive radios

SDR spectrum sensing

- **Analog signal
  conditioning**
- ADC/DAC
  improvements

- Inductor-less
  topologies
- Noise cancelling
  circuits

Wide-band
digitization

Wide-band
circuits

Harmonic
rejection

High
linearity RF

- Harmonic reject
  architectures
- **FFT based
  architectures**

New arch. –
LNTA, mixer first
- Non-linearity
  cancellation

Fig. 6.2 Summary of requirements for signaling in cognitive radios

SDR Spectrum Sensing

1. For the cognitive radio transceiver with its emphasis on flexibility and versatility, the paradigm of performing most functions in the digital domain still remains the popular direction. As a result, the wideband digitization of the RF signal becomes one of the most significant bottle-necks in the realization of cognitive radios.
2. To tackle the wideband digitization problem, analog domain signal conditioning prior to digitization is being considered. Discrete time signal processing is emerging as a popular option.
3. For spectrum sensing, RF circuits and architectures with wideband characteristics, i.e., relatively constant performance (matching, noise-figure) over a large bandwidth are essential.
4. As a result of the large bandwidth of interest, circuit linearity becomes extremely critical. At the receiver, large in-band blockers undergo distortion in non-linear circuits destroying smaller wanted signals (inter-modulation, cross-modulation, gain-compression, etc.).
5. Due to the large bandwidth of operation, the LO harmonics fall within band; consequently, harmonic mixing is a critical issue in both the transmitter and receiver.

Appendix A: SCF

A popular method to extract cyclostationary features is to implement the spectral correlation function (SCF) [65]. In cyclostationary feature detectors, the incoming signal is modeled as a cyclostationary random process with multiple periodicities, as compared to the traditional wide sense stationary model for energy detection where the cyclostationary information is discarded. Then, the autocorrelation function for a zero-mean process $x(t)$ is defined as

$$R_X\left(t+\frac{\tau}{2},\ t-\frac{\tau}{2}\right) = E\left[x\left(t+\frac{\tau}{2}\right)x^*\left(t-\frac{\tau}{2}\right)\right]$$

which, in the 't' domain, exhibits the multiple periodicities of the incoming signal. The Fourier transform of the autocorrelation over 't' as expressed in

$$R_X^\alpha(\tau) = \lim_{Z\to\infty} \frac{1}{Z}\int_{-\frac{Z}{2}}^{\frac{Z}{2}} R_X\left(t+\frac{\tau}{2},\ t-\frac{\tau}{2}\right)e^{-j2\pi\alpha t}dt$$

is called the cyclic autocorrelation in the cycle frequency (α) domain, and is expected to capture these periodicities. For a cycloergodic process, the definition simplifies to

$$R_X^\alpha(\tau) = \lim_{Z\to\infty} \frac{1}{Z}\int_{-\frac{Z}{2}}^{\frac{Z}{2}} x\left(t+\frac{\tau}{2}\right)x^*\left(t-\frac{\tau}{2}\right)e^{-j2\pi\alpha t}dt$$

Then, using the cyclic Wiener-Khinchin relation on $R_X^\alpha(\tau)$, the spectral correlation function (SCF) can then be defined as

$$S_X^\alpha(f) = \int_{-\infty}^{\infty} R_X^\alpha(\tau)e^{-j2\pi f\tau}d\tau$$

Here, we note that the cyclic autocorrelation function, $R_X^\alpha(\tau)$, and the SCF, $S_X^\alpha(f)$, reduce to the conventional definitions of autocorrelation function $R_X(\tau)$, and the power spectral density, $S_X(f)$, for the case when $\alpha = 0$.

Again, an alternative, and more convenient equivalent definition of the SCF can be expressed in terms of the short time Fourier transform (STFT) of the received signal $x(t)$. The STFT of a signal is just the Fourier transform at the particular instance 't', defined as

$$X_T(t,\ f) = \int_{t-\frac{T}{2}}^{t+\frac{T}{2}} x(u)e^{-j2\pi fu}du$$

B. Sadhu, R. Harjani, *Cognitive Radio Receiver Front-Ends*,
Analog Circuits and Signal Processing 115, DOI 10.1007/978-1-4614-9296-2,
© Springer Science+Business Media New York 2014

The SCF can then be defined in terms of the STFT as

$$S_X^\alpha(f) = \lim_{T \to \infty} \lim_{\Delta t \to \infty} \frac{1}{\Delta t} \int_{-\frac{\Delta t}{2}}^{\frac{\Delta t}{2}} \frac{1}{T} X_T \left(t, f + \frac{\alpha}{2} \right) X_T^* \left(t, f - \frac{\alpha}{2} \right) dt$$

thus better justifying its nomenclature. As is evident, this second definition lends itself to an easier execution in hardware and has been used in hardware implementations of cyclostationary feature detectors [113]. For discrete time implementation, the discrete and finite time equivalent definition of the SCF is derived as

$$\hat{S}_X^\alpha(f) = \frac{1}{N} \frac{1}{T} \sum_{n=0}^{N} X_{T_{DFT}} \left(n, f + \frac{\alpha}{2} \right) X_{T_{DFT}}^* \left(n, f - \frac{\alpha}{2} \right)$$

where $X_{T_{DFT}}(n, f)$ represents the N point DFT around sample n.

The superiority of cyclostationary feature detectors over energy detection can be visualized in Fig. A.1, where the power spectral density (PSD) and SCF of a BPSK modulated signal are simulated in MATLAB®. The PSD, as would be calculated by an energy detector is shown in Fig. A.1. The energy detection technique is based on comparing the signal energy with a threshold that is dependent on an estimation of the noise power. As evident in the Fig. A.1, any estimation error due to interference or changing noise variance could easily change the absolute energy detected and cause errors, especially in the case of weak signals. Also, the PSD cannot discriminate between signal energy and noise/interferer energy. However, when the SCF is calculated as in Fig. A.1, the additive white Gaussian noise appears only along the main diagonal, which represents the PSD as estimated by energy detectors. Since the cross-correlation of the noise tends to zero for long observation times, the BPSK signal is clearly visible along the 'α' axis (second diagonal). Moreover, the energy along the 'α' axis is independent of the noise variance.

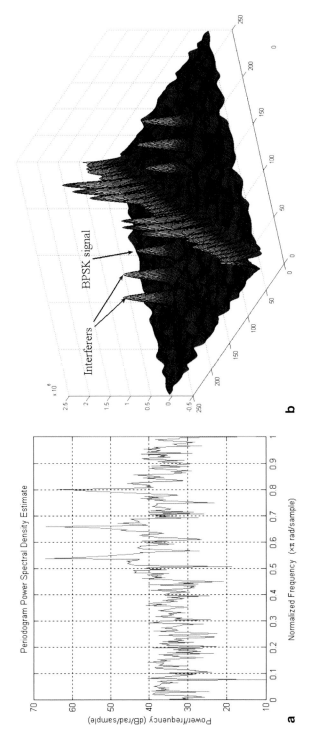

Fig. A.1 (a) PSD of a BPSK signal with − 60 dB SNR and two interferers, and (b) SCF of the same signal, where the diagonal represents the PSD along the f axis as shown in (a)

References

1. Wikipedia, "World cell phone usage," http://en.wikipedia.org/wiki/List_of_countries_by_number_of_mobile_phones_in_use.
2. B. Fette, *Cognitive Radio Technology*, Newnes, 2006.
3. NTIA, "U.S. frequency allocations," http://www.ntia.doc.gov/osmhome/allochrt.pdf.
4. D. Cabric, I.D. O'Donnell, M.S.-W. Chen, and R.W. Brodersen, "Spectrum sharing radios," *IEEE Circuits and Systems Magazine*, vol. 6, no. 2, pp. 30–45, 2006.
5. Federal Communications Commission, "In the matter of facilitating opportunities for exible, effient, and reliable spectrum use employing cognitive radio technologies, report and order," *FCC 05-57, ET Docket No. 03-108*, 2005.
6. T. Ulversoy, "Software defined radio: Challenges and opportunities," *IEEE Communications Surveys Tutorials*, vol. 12, no. 4, pp. 531–550, 2010.
7. J. Mitola III, "Cognitive radio: an integrated agent architecture for software defined radio dissertation," *Royal Institute of Technology*, Stockholm, 2000.
8. J. Mitola III, "Software radios–survey, critical evaluation and future directions," in *Telesystems Conference, 1992. NTC-92., National*, May 1992.
9. R. Bagheri, A. Mirzaei, S. Chehrazi, M. E. Heydari, M. Lee, M. Mikhemar, W. Tang, and A. A. Abidi, "An 800-MHz to 6-GHz software-defined wireless receiver in 90-nm CMOS," *IEEE Journal of Solid-State Circuits*, vol. 41, no. 12, pp. 2860–2876, November 2006.
10. J. Gambini, O. Simeone, Y. Bar-Ness, U. Spagnolini, and T. Yu, "Packet-wise vertical handover for unlicensed multi-standard spectrum access with cognitive radios," *IEEE Transactions on Wireless Communications*, vol. 7, no. 12, pp. 5172–5176, December 2008.
11. O. Gaborieau, S. Mattisson, N. Klemmer, B. Fahs, F.T. Braz, R. Gudmundsson, T. Mattsson, C. Lascaux, C. Trichereau, W. Suter, E. Westesson, and A. Nydahl, "A SAW-less multiband WEDGE receiver," in *IEEE International Solid-State Circuits Conference*, February 2009, pp. 114–115,115a.
12. T. Sowlati, B. Agarwal, J. Cho, T. Obkircher, M. El Said, J. Vasa, B. Ramachandran, M. Kahrizi, E. Dagher, Wei-Hong Chen, M. Vadkerti, G. Taskov, U. Seckin, H. Firouzkouhi, B. Saeidi, H. Akyol, Yunyoung Choi, A. Mahjoob, S. D'Souza, Chieh-Yu Hsieh, D. Guss, D. Shum, D. Badillo, I. Ron, D. Ching, Feng Shi, Yong He, J. Komaili, A. Loke, R. Pullela, E. Pehlivanoglu, H. Zarei, S. Tadjpour, D. Agahi, D. Rozenblit, W. Domino, G. Williams, N. Damavandi, S. Wloczysiak, S. Rajendra, A. Paff, and T. Valencia, "Single-chip multiband WCDMA/HSDPA/HSUPA/EGPRS transceiver with diversity receiver and 3G DigRF interface without SAW filters in transmitter/3G receiver paths," in *IEEE International Solid-State Circuits Conference*, February 2009, pp. 116–117,117a.
13. Sang-June Park, Kok-Yan Lee, and G.M. Rebeiz, "Low-loss 5.15–5.70-GHz RF MEMS switchable filter for wireless LAN applications," *IEEE Transactions on Microwave Theory and Techniques*, vol. 54, no. 11, pp. 3931–3939, November 2006.
14. K. Entesari and G.M. Rebeiz, "A differential 4-bit 6.5-10-GHz RF MEMS tunable filter," *IEEE Transactions on Microwave Theory and Techniques*, vol. 53, no. 3, pp. 1103–1110, March 2005.

15. Chih-Chieh Cheng and G.M. Rebeiz, "High-Q 4–6-GHz suspended stripline RF MEMS tunable filter with bandwidth control," *IEEE Transactions on Microwave Theory and Techniques*, vol. 59, no. 10, pp. 2469–2476, October 2011.

16. G. Rebeiz, K. Entesari, I. Reines, S.-j. Park, M. El-tanani, A. Grichener, and A. Brown, "Tuning in to RF MEMS," *IEEE Microwave Magazine*, vol. 10, no. 6, pp. 55–72, October 2009.

17. R.M. Young, J.D. Adam, C.R. Vale, T.T. Braggins, S.V. Krishnaswamy, C.E. Milton, D.W. Bever, L.G. Chorosinski, Li-Shu Chen, D.E. Crockett, C.B. Freidhoff, S.H. Talisa, E. Capelle, R. Tranchini, J.R. Fende, J.M. Lorthioir, and A.R. Tories, "Low-loss bandpass RF filter using MEMS capacitance switches to achieve a one-octave tuning range and independently variable bandwidth," in *IEEE MTT-S International Microwave Symposium Digest*, June 2003 vol. 3, pp. 1781–1784 vol.3.

18. Yun Zhu, R.W. Mao, and C.S. Tsai, "A varactor and FMR-tuned wideband band-pass filter module with versatile frequency tunability," *IEEE Transactions on Magnetics*, vol. 47, no. 2, pp. 284–288, February 2011.

19. M. Sanchez-Renedo, R. Gomez-Garcia, J.I. Alonso, and C. Briso-Rodriguez, "Tunable combline filter with continuous control of center frequency and bandwidth," *IEEE Transactions on Microwave Theory and Techniques*, vol. 53, no. 1, pp. 191–199, January 2005.

20. J. Nath, D. Ghosh, J.-P. Maria, A.I. Kingon, W. Fathelbab, P.D. Franzon, and M.B. Steer, "An electronically tunable microstrip bandpass filter using thin-film Barium-Strontium-Titanate (BST) varactors," *IEEE Transactions on Microwave Theory and Techniques*, vol. 53, no. 9, pp. 2707–2712, September 2005.

21. Z. Ru, E.A.M. Klumperink, C.E. Saavedra, and B. Nauta, "A 300–800 MHz tunable filter and linearized LNA applied in a low-noise harmonic-rejection RF-sampling receiver," *IEEE Journal of Solid-State Circuits*, vol. 45, no. 5, pp. 967–978, May 2010.

22. L. E. Franks and I. W. Sandberg, "An alternative approach to the realizations of network functions: N-path filter," *Bell Systems Technical Journal*, vol. 59, no. 1, pp. 1321–1350, 1960.

23. A. Mirzaei, H. Darabi, and D. Murphy, "Architectural evolution of integrated M-phase high-Q bandpass filters," *IEEE Transactions on Circuits and Systems I: Regular Papers*, pp. 52–65, January 2012.

24. A. Ghaffari, E.A.M. Klumperink, M.C.M. Soer, and B. Nauta, "Tunable high-Q N-path bandpass filters: Modeling and verification," *IEEE Journal of Solid-State Circuits*, vol. 46, no. 5, pp. 998–1010, May 2011.

25. M.C.M. Soer, E.A.M. Klumperink, P.-T. de Boer, F.E. van Vliet, and B. Nauta, "Unified frequency-domain analysis of switched-series- passive mixers and samplers," *IEEE Transactions on Circuits and Systems I*, vol. 57, no. 10, pp. 2618–2631, October 2010.

26. J.A. Weldon, R.S. Narayanaswami, J.C. Rudell, Li Lin, M. Otsuka, S. Dedieu, Luns Tee, King-Chun Tsai, Cheol-Woong Lee, and P.R. Gray, "A 1.75-GHz highly integrated narrow-band CMOS transmitter with harmonic-rejection mixers," *IEEE Journal of Solid-State Circuits*, vol. 36, no. 12, pp. 2003–2015, December 2001.

27. H. Darabi, "Highly integrated and tunable RF front-ends for reconfigurable multi-band transceivers," in *IEEE Custom Integrated Circuits Conference*, September 2010, pp. 1–8.

28. A. Bevilacqua and A.M. Niknejad, "An ultrawideband CMOS low-noise amplifier for 3.1-10.6-GHz wireless receivers," *IEEE Journal of Solid-State Circuits*, vol. 39, no. 12, pp. 2259–2268, December 2004.

29. A. Ismail and A.A. Abidi, "A 3-10-GHz low-noise amplifier with wideband LC-ladder matching network," *IEEE Journal of Solid-State Circuits*, vol. 39, no. 12, pp. 2269–2277, December 2004.

30. F. Bruccoleri, E.A.M. Klumperink, and B. Nauta, "Generating all two-MOS-transistor amplifiers leads to new wide-band LNAs," *IEEE Journal of Solid-State Circuits*, vol. 36, no. 7, pp. 1032–1040, July 2001.

31. J. Janssens, M. Steyaert, and H. Miyakawa, "A 2.7 volt CMOS broadband low noise amplifier," in *IEEE Symposium on VLSI Circuits*, June 1997, pp. 87–88.
32. F. Bruccoleri, E.A.M. Klumperink, and B. Nauta, "Wide-band CMOS low-noise amplifier exploiting thermal noise canceling," *IEEE Journal of Solid-State Circuits*, vol. 39, no. 2, pp. 275–282, February 2004.
33. F. Tzeng, A. Jahanian, and P. Heydari, "A multiband inductor-reuse CMOS low-noise amplifier," *IEEE Transactions on Circuits and Systems II*, vol. 55, no. 3, pp. 209–213, March 2008.
34. H. Hashemi and A. Hajimiri, "Concurrent multiband low-noise amplifiers-theory, design, and applications," *IEEE Transactions on Microwave Theory and Techniques*, vol. 50, no. 1, pp. 288–301, January 2002.
35. M. El-Nozahi, E. Sanchez-Sinencio, and K. Entesari, "A CMOS low-noise amplifier with reconfigurable input matching network," *IEEE Transactions on Microwave Theory and Techniques*, vol. 57, no. 5, pp. 1054–1062, May 2009.
36. B. Sadhu, J. Kim, and R. Harjani, "A CMOS 3.3-8.4 GHz wide tuning range, low phase noise LC VCO," *Proceedings of IEEE Custom Integrated Circuits Conference*, pp. 559–562, September 2009.
37. B. Razavi. "Cognitive radio design challenges and techniques," *IEEE Journal of Solid-State Circuits*, vol. 45, no. 8, pp. 1542–1553, August 2010.
38. P. Andreani and S. Mattisson, "On the use of MOS varactors in RF VCOs," *IEEE Journal of Solid-State Circuits*, vol. 35, no. 6, pp. 905–910, June 2000.
39. R.L. Bunch and S. Raman, "Large-signal analysis of MOS varactors in CMOS -G_m LC VCOs," *IEEE Journal of Solid-State Circuits*, vol. 38, no. 8, pp. 1325–1332, August 2003.
40. A. Kral, F. Behbahani, and A. A. Abidi, "RF-CMOS oscillators with switched tuning," *Proceedings in IEEE Custom Integrated Circuits Conference*, May 1998.
41. A. D. Berny, A. M. Niknejad, and R. G. Meyer, "A 1.8-GHz LC VCO with 1.3-GHz tuning range and digital amplitude calibration," *IEEE Journal of Solid-State Circuits*, pp. 909–917, April 2005.
42. Z. Ru, N.A. Moseley, E. Klumperink, and B. Nauta, "Digitally enhanced software-defined radio receiver robust to out-of-band interference," *IEEE Journal of Solid-State Circuits*, vol. 44, no. 12, pp. 3359–3375, December 2009.
43. M. Soer, E. Klumperink, Z. Ru, F.E. van Vliet, and B. Nauta, "A 0.2-to-2.0GHz 65nm CMOS receiver without LNA achieving > 11dBm IIP3 and < 6.5dB NF," in *IEEE International Solid-State Circuits Conference*, February 2009, pp. 222–223, 223a.
44. H. Khorramabadi and P.R. Gray, "High-frequency CMOS continuous-time filters," *IEEE Journal of Solid-State Circuits*, vol. 19, no. 6, pp. 939–948, December 1984.
45. J. Silva-Martinez, M. Steyaert, and W. Sansen, *High-performance CMOS continuous time filters*, Kluwer Academic Publishers, 1993.
46. S. Pavan and Y. Tsividis, *High frequency continuous time filters in digital CMOS processes*, Kluwer Academic Publishers, 2000.
47. R. Bagheri, A. Mirzaei, M.E. Heidari, S. Chehrazi, Minjae Lee, M. Mikhemar, W.K. Tang, and A.A. Abidi, " Software-Defined Radio receiver: dream to reality," *IEEE Communications Magazine*, vol. 44, no. 8, pp. 111–118, August 2006.
48. K. Muhammad, Y.C. Ho, T. Mayhugh, C.M. Hung, T. Jung, I. Elahi, C. Lin, I. Deng, C. Fernando, J. Wallberg, S. Vemulapalli, S. Larson, T. Murphy, D. Leipold, P. Cruise, J. Jaehnig, M.C. Lee, R.B. Staszewski, R. Staszewski, and K. Maggio, "A discrete time quad-band GSM/GPRS receiver in a 90nm digital CMOS process," in *IEEE Custom Integrated Circuits Conference*, September 2005, pp. 809–812.
49. K. Muhammad, D. Leipold, B. Staszewski, Y.-C. Ho, C.M. Hung, K. Maggio, C. Fernando, T. Jung, J. Wallberg, J.-S. Koh, S. John, I. Deng, O. Moreira, R. Staszewski, R. Katz, and O. Friedman, "A discrete-time bluetooth receiver in a 0.13 μm digital CMOS process," in *IEEE International Solid-State Circuits Conference*, February 2004, pp. 268–527 vol.1.
50. K. Muhammad, R.B. Staszewski, and D. Leipold, "Digital RF processing: toward low-cost reconfigurable radios," *IEEE Communications Magazine*, vol. 43, no. 8, pp. 105–113, August 2005.

51. R. van de Plassche, "A sigma-delta modulator as an A/D converter," *IEEE Transactions on Circuits and Systems*, vol. 25, no. 7, pp. 510–514, July 1978.

52. M. Bolatkale, L.J. Breems, R. Rutten, and K.A.A. Makinwa, "A 4 GHz continuous-time $\Sigma - \Delta$ ADC with 70 dB DR and -74 dBFS THD in 125 MHz BW," *IEEE Journal of Solid-State Circuits*, vol. 46, no. 12, pp. 2857–2868, December 2011.

53. H. Shibata, R. Schreier, Wenhua Yang, A. Shaikh, D. Paterson, T. Caldwell, D. Alldred, and Ping Wing Lai, "A DC-to-1 GHz tunable RF $\Delta\Sigma$ ADC achieving DR $= 74$dB and BW $= 150$MHz at $f_0 = 450$MHz using 550mW," in *IEEE International Solid-State Circuits Conference (ISSCC)*, February 2012, pp. 150–152.

54. A.M.A. Ali, A. Morgan, C. Dillon, G. Patterson, S. Puckett, P. Bhoraskar, H. Dinc, M. Hensley, R. Stop, S. Bardsley, D. Lattimore, J. Bray, C. Speir, and R. Sneed, "A 16-bit 250-MS/s IF sampling pipelined ADC with background calibration," *IEEE Journal of Solid-State Circuits*, vol. 45, no. 12, pp. 2602–2612, December 2010.

55. S. Devarajan, L. Singer, D. Kelly, S. Decker, A. Kamath, and P. Wilkins, "A 16-bit, 125 MS/s, 385 mW, 78.7 dB SNR CMOS pipeline ADC," *IEEE Journal of Solid-State Circuits*, vol. 44, no. 12, pp. 3305–3313, December 2009.

56. B. Murmann and B.E. Boser, "A 12-bit 75-MS/s pipelined ADC using open-loop residue amplification," *IEEE Journal of Solid-State Circuits*, vol. 38, no. 12, pp. 2040–2050, December 2003.

57. C.C. Lee and M.P. Flynn, "A 12b 50MS/s 3.5mW SAR assisted 2-stage pipeline ADC," in *IEEE Symposium on VLSI Circuits*, June 2010, pp. 239–240.

58. C-C. Liu, S-J. Chang, G-Y. Huang, and Y-Z. Lin, "A 10-bit 50-MS/s SAR ADC with a monotonic capacitor switching procedure," *IEEE Journal of Solid-State Circuits*, vol. 45, no. 4, pp. 731–740, April 2010.

59. Y. Zhu, C-H. Chan, U-F. Chio, S-W. Sin, S-P. U, R.P. Martins, and F. Maloberti, "A 10-bit 100- MS /s reference-freeSAR ADC in 90 nm CMOS," IEEE Journal of Solid-State Circuits, vol. 45, no. 6, pp. 1111–1121, june 2010.

60. S.M. Louwsma, A.J.M. van Tuijl, M. Vertregt, and B. Nauta, "A 1.35 GS/s, 10 b, 175 mW time-interleaved AD converter in 0.13μm CMOS," *IEEE Journal of Solid-State Circuits*, vol. 43, no. 4, pp. 778–786, April 2008.

61. B. Murmann, " ADC performance survey 1997–2012," http://www.stanford.edu/ murmann/ adcsurvey.html.

62. P.B. Kenington, "Linearized transmitters: an enabling technology for software defined radio," *IEEE Communications Magazine*, vol. 40, no. 2, pp. 156–162, Feb 2002.

63. R. Shrestha, E.A.M. Klumperink, E. Mensink, G.J.M. Wienk, and B. Nauta, "A polyphase multipath technique forSoftware-Defined Radio transmitters," *IEEE Journal of Solid-State Circuits*, vol. 41, no. 12, pp. 2681–2692, December 2006.

64. S-J. Kim and G.B. Giannakis, "Sequential and cooperative sensing for multi-channel cognitive radios," *IEEE Transactions on Signal Processing*, vol. 58, no. 8, pp. 4239–4253, August 2010.

65. W.A. Gardner, "Signal interception: a unifying theoretical framework for feature detection," *IEEE Transactions on Communications*, vol. 36, no. 8, pp. 897–906, August 1988.

66. S. M. Mishra, A. Sahai, and R. W. Brodersen, "Cooperative sensing among cognitive radios," *IEEE International Conference on Communications*, pp. 1658–1663, June 2006.

67. Y. Ding and R. Harjani, "A + 18 dBm IIP3 LNA in 0.35 μm CMOS," in *IEEE International Solid-State Circuits Conference*, 2001, pp. 162–163, 443.

68. V. Aparin and L.E. Larson, "Modified derivative superposition method for linearizing FET low-noise amplifiers," *IEEE Transactions on Microwave Theory and Techniques*, vol. 53, no. 2, pp. 571–581, February 2005.

69. W-H. Chen, G. Liu, B. Zdravko, and A.M. Niknejad, "A highly linear broadbandCMOS LNA employing noise and distortion cancellation," *IEEE Journal of Solid-State Circuits*, vol. 43. no. 5, pp. 1164–1176, May 2008.

70. I. D. O'Donnel and R. W. Brodersen, "An ultra-wideband transceiver architecture for low power, low rate, wireless systems," *IEEE Transactions on Vehicular Technology*, pp. 1623–1631, September 2005.

71. S. Hoyos, B.M. Sadler, and G.R. Arce, "Analog to digital conversion of ultra-wideband signals in orthogonal spaces," in *IEEE Conference on Ultra Wideband Systems and Technologies*, November 2003, pp. 47–51.

72. S. Hoyos, B.M. Sadler, and G.R. Arce, "Broadband multicarrier communication receiver based on analog to digital conversion in the frequency domain," *IEEE Transactions on Wireless Communications*, vol. 5, no. 3, pp. 652–661, March 2006.

73. S. R. Valezquez, T. Q. Nguyen, S. R. Broadstone, and J. K. Roberge, "A hybrid filter bank approach to analog-to-digital conversion," *Proceedings of the IEEE-SP International Symposium on Time-Frequency and Time-Scale Analysis*, pp. 116–119, October 1994.

74. W. Namgoong, "A channelized digital ultrawideband receiver," *IEEE Transactions for Wireless Communications*, pp. 502–510, May 2003.

75. T-L Hsieh, P. Kinget, and R. Gharpurey, "A rapid interference detector for ultra wideband radio systems in 0.13μm CMOS," *IEEE Radio Frequency Integrated Circuits Symposium*, pp. 347–350, April 2008.

76. M. Elbadry, B. Sadhu, J. Qiu, and R. Harjani, "Dual channel injection-locked quadrature LO generation for a 4GHz instantaneous bandwidth receiver at 21GHz center frequency," *IEEE Radio Frequency Integrated Circuits Symposium*, June 2012.

77. S. Kalia, M. Elbadry, B. Sadhu, S. Patnaik, J. Qiu, and R. Harjani, "A simple, unified phase noise model for injection-locked oscillators," *IEEE Radio Frequency Integrated Circuits Symposium*, June 2011.

78. B. Sadhu, U. Omole, and R. Harjani, "Modeling and synthesis of wide-band switched-resonators for VCOs," *Proceedings of IEEE Custom Integrated Circuits Conference*, pp. 225–228, September 2008.

79. D. Ham and A. Hajimiri, "Concepts and methods of optimization of integrated LC VCOs," *IEEE Journal of Solid-State Circuits*, pp. 896–909, June 2001.

80. F. Zhang and P.R. Kinget, "Design of components and circuits underneath integrated inductors," *IEEE Journal of Solid-State Circuits*, vol. 41, no. 10, pp. 2265–2271, 2006.

81. J. Craninckx and M. J. Steyaert, "A 1.8-GHz low-phase-noise CMOS VCO using optimized hollow spiral inductors," *IEEE Journal of Solid-State Circuits*, pp. 736–744, May 1997.

82. J. Borremans, S. Bronckers, P. Wambacq, M. Kujik, and J. Craninckx, "A single-inductor dual-band VCO in 0.06mm^2 5.6GHz multi-band front-end in 90nm digital CMOS," Digest of Technical Papers, *IEEE International Solid-State Circuits Conference*, pp. 324–616, Feb 2008.

83. A. Hajimiri and T. H. Lee, "Design issues in CMOS differential LC oscillators," *IEEE Journal of Solid-State Circuits*, pp. 717–724, May 1999.

84. A. Goel and H. Hashemi, "Frequency switching in dual-resonance oscillators," *IEEE Journal of Solid-State Circuits*, pp. 571–582, March 1999.

85. N. H. W. Fong, J. O. Plouchart, N. Zamdmer, D. Liu, L. Wagner, C. Plett, and N. G. Tarr, "Design of wide-band CMOS VCO for multiband wireless LAN applications," *IEEE Journal of Solid-State Circuits*, pp. 1333–1342, August 2003.

86. A. Bevilacqua, F. P. Pavan, C. Sandner, A. Gerosa, and A. Neviani, "Transformer-based dual-mode voltage-controlled oscillators," *IEEE Transactions on Circuits and Systems II*, pp. 293–297, April 2007.

87. Y. Takigawa, H. Ohta, Q. Liu, S. Kurachi, N. Itoh, and T. Yoshimasu, "A 92.6% tuning range VCO utilizing simultaneously controlling of transformers and MOS varactors in 0.13μm CMOS technology," *Proceedings of IEEE Radio Frequency Integrated Circuits Symposium*, pp. 83–86, June 2000.

88. L.R. Carley and T. Mukherjee, "High-speed low-power integrating cmos sample-and-hold amplifier architecture," May 1995, pp. 543–546.

89. A. Mirzaei, S. Chehrazi, R. Bagheri, and A.A. Abidi, "Analysis of first-order anti-aliasing integration sampler," *IEEE Transactions on Circuits and Systems I*, vol. 55, no. 10, pp. 2994–3005, November 2008.

90. G. Xu and J. Yuan, "Comparison of charge sampling and voltage sampling," in *IEEE Midwest Symposium on Circuits and Systems*, 2000, vol. 1, pp. 440–443.

91. S. Karvonen, T.A.D. Riley, and J. Kostamovaara, "Charge-domain FIR sampler with pro-grammable filtering coefficients," *IEEE Transactions on Circuits and Systems II*, vol. 53, no. 3, pp. 192–196, March 2006.

92. J. Yuan, "A charge sampling mixer with embedded filter function for wireless applications," in *International Conference on Microwave and Millimeter Wave Technology*, 2000, pp. 315–318.

93. Z. Ru, E.A.M. Klumperink, and B. Nauta, "On the suitability of discrete-time receivers for Software-Defined Radio," in *IEEE International Symposium on Circuits and Systems*, May 2007, pp. 2522–2525.

94. D. Jakonis, K. Folkesson, J. Dbrowski, P. Eriksson, and C. Svensson, "A 2.4-GHz RF sampling receiver front-end in 0.18-μm CMOS," *IEEE Journal of Solid-State Circuits*, vol. 40, no. 6, pp. 1265–1277, June 2005.

95. R. Bagheri, A. Mirzaei, S. Chehrazi, M.E. Heidari, Minjae Lee, M. Mikhemar, Wai Tang, and A.A. Abidi, "An 800-MHz–6-GHz Software-Defined wireless receiver in 90-nm CMOS,"December 2006, vol. 41, pp. 2860–2876.

96. Z. Ru, E.A.M. Klumperink, and B. Nauta, "Discrete-time mixing receiver architecture for RF-sampling Software-Defined Radio," *IEEE Journal of Solid-State Circuits*, vol. 45, no. 9, pp. 1732–1745, September 2010.

97. A. Abidi, "The path to the Software-Defined Radio receiver," *IEEE Journal of Solid-State Circuits*, pp. 954–966, May 2007.

98. C-H. Heng, M. Gupta, S.-H. Lee, D. Kang, and B-S. Song, "A CMOS TV tuner/demodulator IC with digital image rejection," *IEEE Journal of Solid-State Circuits*, vol. 40, no. 12, pp. 2525–2535, December 2005.

99. V.J. Arkesteijn, E.A.M. Klumperink, and B. Nauta, "Jitter requirements of the sampling clock in software radio receivers," *IEEE Transactions on Circuits and Systems II: Express Briefs*, vol. 53, no. 2, pp. 90–94, February 2006.

100. F. Montaudon, R. Mina, S. Le Tual, L. Joet, D. Saias, R. Hossain, F. Sibille, C. Corre, V. Carrat, E. Chataigner, J. Lajoinie, S. Dedieu, F. Paillardet, and E. Perea, "A scalable 2.4-to-2.7GHz Wi-Fi/WiMAX discrete-time receiver in 65nm CMOS," in *Solid-State Circuits Conference, 2008. ISSCC 2008. Digest of Technical Papers. IEEE International*, February 2008, pp. 362–619.

101. F. Harris, Chris Dick, and Michael Rice, "Digital receivers and transmitters using polyphase filter banks for wireless communications," *IEEE Transactions on Microwave Theory and Techniques*, pp. 1395–1412, April 2003.

102. F. Rivet, Y. Deval, J.-B. Begueret, D. Dallet, P. Cathelin, and D. Belot, "The experimental demonstration of a SASP-based full software radio receiver," *IEEE Journal of Solid-State Circuits*, pp. 979–988, May 2010.

103. M. Lehne and S. Raman, "A 0.13-μm 1-GS/s CMOS discrete-time FFT processor for ultra-wideband OFDM wireless receivers," *IEEE Transactions on Microwave Theory and Techniques*, pp. 1639–1650, November 2000.

104. B. Sadhu, M. Sturm, B. M. Sadler, and R. Harjani, "A 5GS/s 12.2pJ/conv. analog charge-domain FFT for a software defined radio receiver front-end in 65nm CMOS," *IEEE Radio Frequency Integrated Circuits Symposium*, June 2012.

105. B. Murmann, "EE315B: VLSI Data Conversion Circuits," *Stanford University*, 2011.

106. N. J. Guilar, F. Lau, P. J. Hurst, and S. H. Lewis, "A passive switched-capacitor finite-impulse-response equalizer," *IEEE Journal of Solid-State Circuits*, vol. 42, no. 2, pp. 400–409, February 2007.

107. K.W. Martin, "Complex signal processing is not complex," *IEEE Transactions on Circuits and Systems I*, vol. 51, no. 9, pp. 1823–1836, September 2004.

108. B. Sadhu, *Circuit techniques for cognitive radio receiver front-ends*. Dissertation, University of Minnesota, 2012.

109. Y. Ding and R. Harjani, "A universal analytic charge injection model," in *IEEE International Symposium for Circuits and Systems*, 2000, vol. 1, pp. 144–147.

110. G. Wegmann, E.A. Vittoz, and F. Rahali, "Charge injection in analog MOS switches," *IEEE Journal of Solid-State Circuits*, vol. 22, no. 6, pp. 1091–1097, December 1987.

111. M.C.M. Soer, E.A.M. Klumperink, P.-T. de Boer, F.E. van Vliet, and B. Nauta, "Unified frequency-domain analysis of switched-series passive mixers and samplers," *IEEE Transactions on Circuits and Systems I*, vol. 57, no. 10, pp. 2618–2631, October 2010.
112. Y.-W. Lin, H.-Y. Liu, and C.-Y. Lee, " A 1-GS/s FFT/IFFT processor for UWB applications," *IEEE Journal of Solid-State Circuits*, pp. 1726–1735, August 2005.
113. A. Tkachenko, A.D. Cabric, and R.W. Brodersen, "Cyclostationary feature detector experiments using reconfigurable BEE2," in *IEEE International Symposium on New Frontiers in Dynamic Spectrum Access Networks*, April 2007, pp. 216–219.

Printed by Publishers' Graphics LLC
LMO140107.15.16.58